# SLEEPYHEAD

ALSO BY HENRY NICHOLLS

*The Galapagos*
*The Way of the Panda*
*Lonesome George*

# SLEEPYHEAD

The Neuroscience of a
Good Night's Rest

## HENRY NICHOLLS

BASIC BOOKS
*New York*

Basic Books
Hachette Book Group
1290 Avenue of the Americas, New York, NY 10104
www.basicbooks.com
Printed in the United States of America

Originally published in hardcover and ebook by Profile Books in the United Kingdom in March 2018

First U.S. Edition: September 2018

Published by Basic Books, an imprint of Perseus Books, LLC, a subsidiary of Hachette Book Group, Inc. The Basic Books name and logo is a trademark of the Hachette Book Group.

The Hachette Speakers Bureau provides a wide range of authors for speaking events. To find out more, go to www.hachettespeakersbureau .com or call (866) 376-6591.

The publisher is not responsible for websites (or their content) that are not owned by the publisher.

Library of Congress Control Number: 2018939612

ISBNs: 978-1-5416-7257-4 (hardcover); 978-1-5416-7256-7 (ebook)

LSC-C

10  9  8  7  6  5  4  3  2  1

To Charlotte

# Contents

# Preface

This was not meant to be a self-help book. I certainly never imagined it would help me as it has done.

When I first began to think about this project, I considered writing a book solely about narcolepsy (a sleep disorder that I have lived with for more than 20 years). My agent and publisher encouraged me to go further, to cover the whole of sleep and many sleep disorders. I understood their thinking. The bigger, broader book might have bigger, broader appeal. It's probably safe to say that nobody was thinking that embarking on a more wide-ranging project would improve my own sleep. But it has – immeasurably.

By surveying a wide range of sleep pathologies, I have come to see narcolepsy in a very different light, not as an isolated sleep disorder but one with real and important connections to just about every other sleep problem out there. As I've learned about bad sleep in all its many forms, so I've come to appreciate what good sleep means and how to achieve it. This revelation, I believe, has important implications, not just for those with narcolepsy but for everyone who wants to improve their sleep.

I now have such good sleep, more often than not startled afresh by my new-found ability to function, that this project has already been worth its while. But authors write to be read and I hope that this book and its message will reach the wider audience envisaged. In fact, this book is really for anyone who wants to know more about sleep and why it's so very important.

Before we begin, I should just acknowledge that many people with narcolepsy understandably rail against being referred to as 'narcoleptic', as if they are defined by the condition. The same goes for people with insomnia and the 'insomniac' label. I tried hard to avoid these adjectives, but I found it impossible. Forgive me. I use narcoleptic and insomniac here for their literary expediency and on the understanding that they are shorthand for the more appropriate but clunky terms 'person with narcolepsy' and 'person with insomnia' or their unutterable abbreviations pwn and pwi.

# Bad sleep

*'Sleep, those little slices of death; Oh how I loathe them.'*
Edgar Allan Poe

'How is your necrophilia?'

My friend was asking after my narcolepsy, a seriously disabling neurological condition that I have lived with now for more than half my life.

'It's still an issue,' I replied. 'Thanks for asking.'

'Narcolepsy', roughly translated from ancient Greek, means 'an attack of sleep', a reference to its key symptom of excessive daytime sleepiness. But the passing resemblance of the word 'narcolepsy' to 'necrophilia' and 'nymphomania' can result in some awkward conversations. In fact, this mistake is sufficiently common for one person with narcolepsy to have worked up a couple of useful aphorisms. Necrophilia is sleeping with the dead, whereas narcolepsy is being dead asleep. Nymphomania is the urge to sleep with lots, whilst narcolepsy involves lots of sleep.

The onset of most sleep disorders is usually slow, so slow it can be difficult to notice what is happening. I was 21 and halfway through my second year at university when I experienced my first symptoms. I had always been an early riser. I loved mornings, eager for the day ahead. But gradually, over the space of several months, something began to change.

Within an hour of waking, it felt as if a smog were seeping into my brain and anaesthetising my faculties until I was left with no option but sleep. If circumstances allowed, I would give in and go back to bed, but I soon found that more sleep did nothing to refresh me. I tried to fight it instead, for a while – going to lectures, tutorials, running the college bar – but the constant struggle to prevent myself from drowning was almost worse.

I came up with little tricks that would hold off sleep for a minute or two, pinching myself, running violently on the spot, shouting at the top of my voice. Just writing about this makes me feel a little nauseous, even now. The feeling would begin to creep up on me all over again. I could battle like this for tens of minutes, convinced I was managing to take in a book, watch a film or even chat with friends, only to find I had no recollection of what had just happened. Had I been asleep, or hadn't I? The alternatives seemed to be stark: either spend all day in this zombie-state, or simply stay in bed. I began to confront the fact that I couldn't function in a meaningful way. I struggled through my degree, dribbling over lecture notes, dozing through tutorials, overdosing on caffeine to extend the fitful snatches of wakefulness I needed to make the grade.

In some lectures, I was just one of some 500 students. With tiers of desks stretching up and far away from the lecturer, I could sit near the back and put my head down without fear of being spotted. In a tutorial, however, with one professor and two students, things were not so easy. Very quickly, often within minutes, the small, sometimes windowless room and hulking radiators would conspire to bring on an attack of sleep. In order to keep up appearances, I had to keep my eyelids open. With considerable effort I found I could do this

2

but I know (because I've seen it in other people with narco-lepsy) that there would have been no spark, no life, in my eyes.

\* \* \*

■ *Sleep v.*
*To take repose by the natural suspension of consciousness*

It is perhaps understandable that sleep, as defined here by the *Oxford English Dictionary* and as most of us understand it, is a fairly vague concept. Understandable because it is so extraordinarily variable, differing wildly from one conscious species to the next.

Even within mammals – relatively recently evolved with very little variation in central sleep-regulating circuitry of the brain – there are many different takes on sleep. The giraffe, for instance, kips for no more than five hours a day. At the other end of the spectrum is the American opossum, a species that according to a student investigation at Yale University School of Medicine in the 1960s, may sleep for as much as 20 hours out of 24.

In between these extremes, there is a remarkable diversity of sleep. Most mammals are polyphasic in their sleep, drifting off repeatedly over the course of 24 hours. Dogs are the perfect exponents of this approach. Some species are biphasic, with two discrete sleeps a day. Rabbits, for instance, are most active at dawn and dusk, a crepuscular existence that carves out two clear periods for rest, one in darkness and one in daylight. Marine mammals, like dolphins and whales, boast the remarkable ability to sleep with one half of the brain at a time. Then there is also a handful of species with monophasic sleep, dividing up each day into one clear period of wakefulness and one

3

period of sleep. Some of these are nocturnal in their habits, like owls, some are diurnal, like people.

Humans have evolved to sleep more than a giraffe and less than an opossum. We like to sleep during the hours of darkness. We are a species that seems to need somewhere between six and eight hours of sleep in a 24-hour period. But even within a single species there is tremendous variation and sleep can vary considerably from one person to the next. Our genetics, gender, age, the seasons and our cultural traditions, not to mention the whims of everyday life, all have an effect on how we sleep.

It is only relatively recently that we have started to pay attention to this variation. For much of human history, sleep was seen as an inconvenient distraction from everything that goes on during the hours of wakefulness. 'Sleep is evidently a privation of waking,' wrote Aristotle in the fourth century BC, drawing an analogy between sickness and health, ugliness and beauty, weakness and strength. In the Old Testament, sleep is deployed as a metaphor for death so frequently it's easy to start thinking that just as wakefulness follows sleep, life might follow death and that crucifixion could be followed by resurrection. Poets too have often referred to sleep rather than death, using its reversible status to soften the dark finality that is the end of life.

Over two millennia after Aristotle, it was not uncommon to find the medical profession casting similar aspersions on sleep. 'The alternation between watchfulness and sleep ... has its origin in the imperfection of our nature,' Dr Wilson Philip told the Royal Society in 1833. Sleep, if not quite 'a positive evil', is clearly a 'defect', he argued.

There are plenty of people who dismiss sleep as if it's some bothersome encumbrance. When asked how much sleep people needed, Napoleon Bonaparte is fabled to have said: 'Six for a man, seven for a woman, eight for a fool.' Margaret Thatcher is thought to have got by nicely on only four hours a night. More recently, Donald Trump is an advocate of cutting back on kip. 'Don't sleep any more than you have to,' he wrote in his 2004 best-seller *Think Like a Billionaire*. 'I'm not a big sleeper,' he reiterated during his presidential campaign in 2016. 'I like three hours, four hours. I toss, I turn, I beep-de-beep, I want to find out what's going on.' These voices, prominent as they might be, should not be listened to. They ignore three different bodies of evidence.

First, there is the plain and simple, gut-instinctive wisdom that sleep is a force for good rather than evil. Shakespeare, for instance, had Macbeth describe sleep as 'sore labour's bath, Balm of hurt minds, great nature's second course, Chief nourisher in life's feast'. Miguel de Cervantes was similarly insightful, with Sancho Panza telling Don Quixote that sleep is 'meat for the hungry, drink for the thirsty, heat for the cold, and cold for the hot. It is the current coin that purchases all the pleasures of the world cheap, and the balance that sets the king and the shepherd, the fool and the wise man, even.' Samuel Taylor Coleridge was also upbeat in *The Rime of the Ancient Mariner*, characterising sleep as 'a gentle thing, Beloved from pole to pole!'

Then there are the data, a body of scientific observation and experiment that has been steadily growing for around a century. 'Far from being the opposite of wakefulness, sleep is in reality a complement to the waking state, the two constituting alternative phases of a cycle, the one related to the other as the trough of a wave is related to the crest,' wrote sleep science pioneer Nathaniel Kleitman in his groundbreaking book *Sleep and Wakefulness* published in 1939.

Finally, there are the stories of those who live with a sleep disorder. These are often anecdotal in nature, so don't often cut the mustard as scientific evidence and are not often heard. Yet the testimony of those with bad sleep is compelling. When sleep goes wrong, it almost always has a profound effect on our lives, eroding both mental and physical health and causing serious psychological damage. It is time to stop ignoring these first-hand accounts.

There remains plenty of debate over what is going on when we sleep and what possible function sleep might serve. Perhaps our brains spend the downtime clearing out the by-products of metabolism that accumulate during our waking, thinking hours. Maybe the brain cells are taking time out to perform some operation, like consolidating memory or expunging unwanted cognitive clutter. Sleep could simply be a cost-saving strategy, a way of passing time without expending too much energy. Alternatively, sleep could have a role in all of the above, and more.

For now it's sufficient to reiterate that sleep is the result of an evolutionary process, sleep is a widespread phenomenon and the sleep of one species is very different from the sleep of the next. Given these observations, it's safe to conclude that sleep must serve an absolutely vital function. If it doesn't, as sleep researcher Allan Rechtschaffen put it so nicely in 1971, 'then it is the biggest mistake the evolutionary process has ever made'.

\* \* \*

The most obvious way to interfere with this evolutionary imperative is to alter the duration of sleep. As sleeping less than six hours a day is easily arranged, it is understandable that the majority of research should focus on what happens when we don't sleep enough. This demonstrates that too little sleep is associated with a long list of undesirable health outcomes, like an increase in appetite, risk of obesity, high blood pressure, susceptibility to infection, likelihood of depression and pace of cognitive decline.

In addition to the quantity of sleep, the quality of the sleep that we get is tremendously important too. One hundred years ago, most people who'd given it a moment's thought assumed that the brain wasn't up to much during the hours of sleep. In the 1930s, scientists discovered that the electronic signals pulsing over the scalp go through a series of clear changes throughout the night. The precise order and duration of these stages can make a huge difference, so a short but well-structured sleep could leave you feeling more rested than a long but poorly structured one.

This mention of sleep structure (or 'sleep architecture' as it's known in the trade) invites a useful analogy. The perfect night's sleep is like the Taj Mahal, a construction whose beauty emerges from its carefully proportioned design and attention to detail. When the structural integrity of sleep goes awry, as occurs in most sleep disorders, it's as though the architectural plans have gone missing and the resulting edifice is simply a miscellaneous jumble of marble.

With sleep varying so wildly from person to person, from one night to the next, and its timing, duration and structure influenced by so many factors, such as late nights, alcohol, caffeine and stressful deadlines, there are plenty of everyday explanations for bad sleep. This means that it can take many months, often many years, sometimes even decades before someone with a genuine problem plucks up the confidence to take their symptoms to a doctor.

For Monsieur 'G', a 38-year-old wine-barrel retailer from

Paris, the time came on 15 February 1879. His doctor, Jean-Baptiste Édouard Gélineau, listened carefully to his patient, noting how he'd experience bouts of 'deep heaviness' and 'a heavy load on the forehead and deep in the eyes', and felt that there could be something serious going on here, a genuine sleep disorder that needed a name of its own. So Gélineau coined the term 'narcolepsy' and used a delightful Gallic turn of phrase to sum up the principle symptom of this newly described condition: 'an invincible need for sleep'.

But what did Gélineau know? He was only a provincial doctor from Rochefort and the medical elite in Paris dismissed the case out of hand. Perhaps they took one look at his patient's profession – barrel retailer – and figured that the more likely explanation for his sleepiness was inebriation. Whatever the reason for their rejection of Gélineau and patient 'G', it meant narcolepsy did not become an accepted condition until well into the twentieth century.

Even then, the experience of being bounced from one unfruitful GP appointment to the next is all too common for people with dysfunctional sleep.

\* \* \*

Dorothy Ennis-Hand developed narcolepsy in 1953 at the age of 16 and recalls the first time she went to a doctor to get help. She'd left her home in Dublin to take up a factory job at Joseph Lucas Ltd in Birmingham, a major manufacturer of engine parts. Finding herself struggling with sleep in the drawing office, she went to a doctor, he suggested it was probably a result of anxiety and prescribed a bottle of a mysterious 'tonic' as a pick-me-up. 'It didn't help,' Dorothy tells me. A year or so later, by this time living in Ottawa, she sought a second opinion. The Canadian doctor didn't do much better, latching on to the revelation that she'd just broken up with a boyfriend. This was probably the explanation for her sleepiness, he told her. A few years later and back in Dublin, Dorothy made yet another doctor's appointment, only to be met with more speculation. 'Sometimes newborn babies get their day and night mixed up,' he told her, implying that she must be experiencing something similar. His answer was to prescribe two drugs, one to keep Dorothy awake in the daytime and one to put her to sleep at night. Unfortunately, the sedative left her like a zombie. At work, as an orderly in the operating theatres of the Rotunda Hospital, her fellow nurses assumed she'd been drinking over the weekend. But Dorothy – a teetotaller – knew better. 'I threw all the medicine down the loo,' she says.

Several years later, after a series of bitty jobs in Dublin, Los Angeles and Liverpool, and a short-lived marriage, Dorothy found herself back in England, working full-time in the operating theatres at Manchester Infirmary and looking after her young daughter and her youngest brother. 'It was a very stressful time,' she remembers, and the sleep had become more of an issue than ever. Finally, she managed to get a diagnosis. By a stroke of good fortune, the doctor she saw had come across another case of narcolepsy and referred her to a neurologist who made the formal diagnosis. Dorothy was 33 and had lived with narcolepsy – undiagnosed – for just over 15 years.

It's easy to imagine that the repeated failings to diagnose Dorothy's narcolepsy – in Birmingham, Ottawa and Dublin – were because she was born in the 1930s and presenting symptoms to her doctors in the 1950s at a time when narcolepsy was not on the medical radar. Yet the same kind of thing was still happening decades later, in the 1980s.

Michelle Hicks was born in 1975 and was hit by narcolepsy at the unusually young age of seven. Sleeping at primary school quickly became an issue for Michelle. 'I used to go into quite deep sleeps,' she says. 'I'd just zonk out through the whole lesson.' On more than one occasion, she remembers waking up to an empty classroom. 'Everyone had gone home from school.' Her mother and sometimes her teacher would be there with her, sitting and waiting for her to come round.

Alarmed, Michelle's parents took her to the doctor and, at the age of seven, she spent a week in the Whittington Hospital in north London while they ran some tests. 'I remember them giving me blood tests every day and doing x-rays.' By the end of the week, the doctors couldn't find anything wrong with her, but they did refer her to a psychiatrist.

Michelle has the letters that the psychiatrist wrote to her GP in which he seemed more interested in the stress that she was causing her parents. He suggested the family adopt a 'bedtime ritual', getting together and teaching Michelle how to dance. 'This would have the function of bringing mother and father together more in something which they could enjoy together as a couple.' He arranged to see them all again in a month's time. 'I feel quite optimistic that I will be able to help, both with Michelle's symptoms, fears and anxieties and in relation to the parents' feelings of depression and sadness.' The dancing didn't help.

Then, in the early 1990s, when Michelle was in her late teens and had been living with undiagnosed narcolepsy for more than a decade, she was listening to a radio phone-in on Capital Gold and the listeners were sharing their stories of sleeping in unusual places. A woman called to say that her husband suffered from narcolepsy and fell asleep all over the place: while writing, in the car waiting for the traffic lights to turn green. It was the first time Michelle had ever heard the word 'narcolepsy'. 'That's me,' she thought.

Michelle got her hands on a leaflet and her father spoke to someone at the UK Association for Narcolepsy, a charity that Dorothy Ennis-Hand had recently founded that subsequently evolved into Narcolepsy UK. Empowered with what appeared to be an accurate self-diagnosis, she went to her GP, who referred her to a neurologist, who in turn ruled out narcolepsy on the basis that the symptoms had appeared at such an early age. 'We've only seen it in adolescents,' she told Michelle. 'You're anaemic and suffer from depression and that makes you tired.' By this time, Michelle had lived with narcolepsy – undiagnosed – for most of her life and to an extent she had come to accept a life of sleep.

It took another ten years before Michelle had built up the confidence to seek another opinion. It was 2004 and she was studying for a foundation degree at Farnborough College of Technology. Her GP took Michelle's self-diagnosis seriously, referring her to a specialist at Frimley Park Hospital in Surrey, a different hospital than the one she'd visited in the 1990s. When she entered the consulting room, she was dismayed to find herself face to face with the same neurologist she'd seen a decade earlier, recently moved to Frimley Park.

'She thought there were signs of narcolepsy and sleep apnea,' remembers Michelle, but sent her to be tested for the apnea first. When the results came back negative, Michelle still had no explanation for her sleepiness and did not relish a follow-up appointment with the same doctor. Consequently, it was only in 2010, when Michelle was 35 and had battled her symptoms – alone – for almost three decades, that she finally got a proper diagnosis.

* * *

Here's the problem. Until recently, medical students have received next to no training on sleep disorders and the importance of sleep. In the 1990s, researchers in the US sent out a survey to over 125 medical schools across the country. This revealed that over the course of a seven-year training programme, a typical medical student in America would receive less than two hours of training on sleep and sleep disorders. Indeed, around one quarter of all medical schools admitted that they had absolutely no training in this area at all. The authors of the paper – one of them a student of Nathaniel Kleitman's called William Dement (from whom we'll hear much more in due course) – came to a dismal conclusion: 'It appears that physician education in sleep and sleep disorders is largely inadequate, despite increasing evidence of the role of sleep in patient health and well-being.'

A similar study in the UK a few years later made for even starker reading, with undergraduate medics getting around five minutes on sleep over the course of a seven-year degree. Almost 20 years after its publication, the author of the paper, Gregory Stores, now emeritus professor of developmental neuropsychiatry at the University of Oxford, says: 'I fear not much has changed in the meantime. Improvement is still needed all round.'

This view is underscored by a recent survey in the US, which sought to get to grips with what medical professionals know about narcolepsy. Only one in four primary care physicians and two in three sleep specialists considered themselves 'very' or 'extremely' knowledgeable about narcolepsy. Only one in ten of the primary care physicians and less than half of the specialists felt 'very' or 'extremely' comfortable diagnosing the disorder. I am not sure I have ever met a single person with narcolepsy for whom the route to diagnosis was straight-forward.

Even if a doctor does not dismiss a sleepy patient but takes their symptoms seriously, it is not uncommon to come away with a misdiagnosis. A few years ago, Meir Kryger, a sleep specialist at Yale School of Medicine and the author of *The Mystery of Sleep*, hit upon a neat idea. By combing back through medical records, he and a colleague found that in the year prior to a formal diagnosis, patients with narcolepsy typically make around ten visits to the doctor (roughly twice as many as normal). Where there was no diagnosis of narcolepsy (in around two out of every three cases), the doctors came to various miscellaneous conclusions, either just describing the symptoms or suspecting either epilepsy or some kind of mental health disorder like depression.

For many other sleep disorders, the situation is likely to be considerably worse, because narcolepsy, although it only affects around one in 2,500 people, is relatively well understood. There is idiopathic hypersomnia, for instance, the 'idiopathic' label a frank admission that nobody has the foggiest what is going on, and IHers (as they refer to themselves) can find it very much harder to be taken seriously. Likewise Kleine-Levin Syndrome, also known as 'sleeping beauty syndrome', in which the victim – usually in their teens – falls into a sleep-like state for weeks and months on end. Although the symptoms are striking, it's so rare – affecting just one in a million or so – that few doctors have the knowledge to make a diagnosis.

The wide natural variation in patterns of normal sleep, the unpushiness of patients (because there are often other plausible explanations) and the lack of medical training about sleep disorders conspire to create an exhaustingly long road to diagnosis. For reasons we will explore in due course, narcolepsy is rarely present at birth but tends to develop during adolescence, peaking at around 15 years old. The age of diagnosis charts a similar pattern, only smeared into a much wider distribution, peaking at around age 30. Indeed, most studies that have looked at the delay between the onset of symptoms and diagnosis report an average wait of around 15 years.

It felt like forever at the time, but I only had to battle for 18 months until I got a diagnosis. This makes me one of the lucky ones. Dorothy's experience is fairly typical. For Michelle, the delay between onset of symptoms and diagnosis was well above average. Some bear the burden of undiagnosed narcolepsy for a lifetime. In a study of Europeans, the longest recorded wait from onset of symptoms to diagnosis is a remarkable 67 years.

\* \* \*

In recent years, there has been increasing awareness of the importance of sleep and sleep disorders like narcolepsy, and access to healthcare advice, blogs and videos, though the internet has accelerated the journey to diagnosis considerably. Yet in spite of these welcome improvements, there needs to be a far better understanding of what bad sleep looks like.

In the case of narcolepsy, for instance, I am thrilled to find that most people I talk to about it know that it has something to do with too much sleep. It is true that people with narcolepsy can fall asleep in surprising places: at huge sporting events, rock concerts or next to speakers at nightclubs; at the hairdressers; in the tattoo parlour; sitting at the bus stop; on a bench in the city centre in the week before Christmas; in a small boat sailing around the Farne Islands, with the freezing North Sea cascading over the gunwale; while scuba diving, at a depth of 20 metres below the surface; on a rollercoaster;

during a root canal operation at the dentists; on the back of a horse; in a magnetic resonance imagery machine; playing Bingo; on top of a mountain in North Africa; on a surfboard. Though most examples are more mundane.

This cannot be said for the other, lesser-known symptoms that characterise narcolepsy. These can be devastating, often more so than the excessive sleep. But these add-on symptoms are extremely interesting too, peculiar pathologies that give a rare insight into many other aspects of dysfunctional sleep.

A few years ago, for instance, I was in a ski resort in the French Alps. It was late in the season and spring was in the air. I was on a chair lift passing over a tired slope from which the snow had recently disappeared. Rivulets of melt trickled between rocks and tufts of grass. I was scouring the terrain, in the vague hope of spotting a squirrel-like marmot that might have emerged from hibernation when – lo and behold – I saw one.

'A marmot!' I said to my friend Kate sitting next to me on the lift, only the words came out as a slur, my head dropped to my chest and my skis fell from the footrest and began to tug at my collapsing torso. I would have slipped out from the padded seat, but my body became wedged in by my ski poles, preventing both them and me from falling the ten or so metres to the muddy earth below.

This is cataplexy, in which intense emotion causes all the muscles around the body to cut out, just as occurs during one of the stages of sleep. Because it's most often triggered by positive emotions like mirth and elation, occurs in a swirl of good humour and is usually over in a matter of seconds, it is easy to imagine that cataplexy might not have that much of an effect on everyday life. But, as we will see, this is not true at all. Like the sleepiness of narcolepsy, cataplexy works its dark magic slowly.

For decades, sleep specialists have been referring to 'the tetrad of narcolepsy', four key symptoms that are present in different combinations in different people. In addition to excessive daytime sleepiness, there is also often cataplexy, sleep paralysis and terrifying nocturnal visions or hypnagogic hallucinations. I am not unusual in suffering from all four elements of the narcoleptic tetrad.

More recently, there have been calls to transform the tetrad into a pentad, the fifth and somewhat paradoxical symptom being disturbed, incredibly fragmented night-time sleep. But why stop with a pentad? It is becoming increasingly clear to me that there are very real connections between narcolepsy and just about every other sleep disorder you care to mention, including circadian sleep disorders, sleep-disordered breathing, strange automatisms like sleep-related eating disorder, movement disorders such as periodic limb movement disorder and even insomnia.

In order to make sense of sleep and the many ways it can go wrong, it's necessary to understand a bit about the absolutely fundamental, far more evolutionarily ancient role that light plays in dividing every day into periods of activity and inactivity.

## 2

# Let there be light

*'Light affects our circadian rhythms more powerfully than any drug.'*
Charles Czeisler

We are addicted to light.

When earth's very first lifeforms appeared some 3.8 billion years ago, the planet was spinning at a heck of a pace. So fast, in fact, that according to some estimates, the sun would have been whipping across the sky in just 30 minutes, rising afresh every hour or so. This perpetual rhythm – light and dark, hot and cold – has wired the chemistry of life.

The first clear indication that this had to be so came almost 300 years ago, when the French astronomer Jean Jacques d'Ortous de Mairan stuck a plant in a closet. It was a *Mimosa*, most likely *M. pudica*, a species variously referred to as the shameplant or touch-me-not owing to its shrinking response to contact. De Mairan was intrigued by the way the *Mimosa* leaves, each a feathery array of smaller leaflets, closed up at night and opened again during the day. If, as seemed likely, this movement were orchestrated by the sun, a cupboard-confined *Mimosa* should simply close up and stay closed. To de Mairan's surprise, the leaflets continued to rise and fall even in complete darkness, leading him to conclude that the plant was somehow able 'to sense the Sun without ever seeing it'.

It was only centuries later, in the 1970s, that the mechanism behind this remarkable phenomenon began to emerge. Researchers studying lab-bred, mutated fruit flies noticed that a few of them had unusual patterns of activity. While most had a regular 24-hour rhythm to their lives, one strain seemed to operate on a shorter 19-hour cycle, one on a drawn-out 28-hour cycle and another had no discernible cycle at all. When researchers located the mutations responsible, they were all in the same gene. It seemed reasonable to conclude that it had some crucial role in setting up these oscillations and they called it *Period*. More genes followed – *Timeless, Clock, Double-Time* – and then a long list of others with less obvious names like Brain Muscle Aryl Hydrocarbon Receptor Nuclear Translocator-like 1, or *Bmal1* for short.

Every living cell that has ever been studied has its own suite of pacemaker genes. Many of these networks have evolved independently in different branches of the tree of life, but the sequence of the genes and the role they play in the network is often remarkably similar, even in distantly related species. The *Period* gene, for instance, is not confined to fruit flies, but versions of it are found in almost all mammals, including humans. It is these networks that explain de Mairan's observations on *Mimosa*. Within each cell of every living organism, there is an internal, never-ending cycle of molecular activity that is in very close agreement with earth's axial rotation.

This is not magic. It does not support the existence of some cosmic being with a penchant for intelligent design. Rather, the coincidence between these cellular circuits and day length is one of the finest illustrations of the power of natural selection. The sun is so fundamental to life on earth that there is intense pressure on cells to synchronise with our nearest star. Any cellular or multicellular organisms that do not dance to its tune tend not to survive and reproduce as well as those that do.

This molecular circuitry is the basis of what is commonly referred to as the biological clock. This clock keeps 'circadian' time, an elision of the Latin words *circa* (about) and *diem* (a day). The circadian rhythm plays an orchestrating role in every cell-based organism on earth, including humans. In 2017, Jeffrey Hall, Michael Rosbash and Michael Young received the Nobel Prize for Physiology or Medicine for their discoveries that revealed the molecular circuitry responsible for the circadian rhythm within each cell.

\* \* \*

I understand how a simple single-celled organism like a bacterium or an amoeba might be able to mete out a circadian rhythm. What I find harder to comprehend is how a multicellular being as complex as a human manages to express a coherent circadian rhythm. An adult human body contains some 40 trillion cells, each with its own cycle of molecular activity. Keeping these to time is going to require some serious coordination. In animals, this is the job of the suprachiasmatic nucleus, or SCN, twinned clusters of cells in the hypothalamus, a tiny and incredibly influential structure in the centre of the brain that is not just involved in the regulation of the circadian rhythm, but also hunger, thirst, sexual behaviour, body temperature, the daily flip-flop between sleep and wakefulness and much more besides.

There is a video on YouTube (search for '32 Metronome Synchronization') that offers an analogy for how the SCN works its orchestrating wizardry. The clip opens with the 32 metronomes arranged in four rows of eight on a flat, horizontal surface. Some are orange, some are pink, some are green. A hand appears above the back row and taps the metronomes into motion. Even though they are all set to the same speed, they swing out of synch, the cacophony of near-constant noise a perfect expression of chaos.

Then something remarkable happens because the surface on which the metronomes are standing is suspended from the ceiling. As it moves in response to the toing and froing, all these wildly asynchronous metronomes are slowly drawn into a single, perfect left-right rhythm, the coordinated clicks sounding a lot like the in-step boots of a well-drilled army. This offers a clever analogy for the way the SCN underpins the cellular clocks all round the body and marshals them into perfect line.

The central coordinating role of the SCN was confirmed by circadian researchers at the University of Oregon, who described a mutation in golden hamsters that dramatically sped up the molecular machinery within each cell. The affected rodents experienced the ebb and flow of the circadian tide once every 20 hours rather than the normal, near-24-hour period. Subsequently, by transplanting SCN cells from the mutants, researchers were able to transform normal hamsters, so that every single cell in their bodies danced to the new, 20-hour tune.

When the SCN is running to time, perfectly aligned with the 24-hour period of the sun, one of its many functions is to prevent sleepiness from making an inappropriate appearance. With every minute of wakefulness, the body's need for sleep builds. In the middle of the afternoon, many people will be aware of this so-called 'sleep pressure' as it threatens to take over, but then sleepiness vanishes. This is down to the SCN, which delivers a powerful alerting signal at around this time, counteracting the 'sleep pressure' and keeping sleep at bay until the late evening.

The way the SCN achieves this brain-boosting function and orchestrates activity throughout the rest of the body is likely to be fiendishly complex, acted out through the interplay between many neurotransmitters, neuromodulators, hormones, metabolites and the rise and fall of body temperature. While researchers try to unravel this complexity, they have found that one molecule, the hormone melatonin, happens to be a very good marker of where the SCN is in its molecular cycle and have used this to gain all sorts of interesting insights.

Like any biological trait, there is natural variation in the SCN pacemaker (as measured by the cycle of melatonin concentration). Some beat faster than others: the SCN period of the average human is 24 hours and 12 minutes rather than 24 hours exactly. In order to correct this deviation from 24, the SCN relies on what circadian biologists refer to as 'zeitgebers', literally time-givers. Owing to the earth's axial rotation, the sun gives off several cues that change reliably on a 24-hour basis, like light, temperature and humidity, for instance, all of which the SCN can use to keep to a 24-hour cycle. Of these cues, the most powerful signal – by far – is light.

When photons enter our eyes, our brains construct an image of the world around us. But the eyes are about much more than sight, says Jamie Zeitzer, a circadian biologist at Stanford University in California with a particular interest in the way that light affects the central rhythm of the SCN. In addition to the rods and cones that we use to see, the retina also contains other light-receptive cells that send messages directly to the SCN.

The SCN is particularly sensitive to blue light, and this is no coincidence. Owing to the way that light is refracted when it hits the earth's atmosphere, blue wavelengths begin to reach us before sunrise and can still be seen after sunset. So by paying special attention to the blues, the SCN is making a mental note, quite literally, of sunrise and sunset.

The light that enters the eyes between dawn and dusk is important too. Inside Zeitzer's office, where the blinds are drawn, he estimates that I am probably only receiving about 300 lux of light. Zeitzer, who has his back to the window, is getting less still. Outside Zeitzer's office, where the sun is shining brightly, he estimates I'd probably be getting around 50,000 lux. This kind of exposure is important because without it, the SCN becomes overly sensitive to the twilight blues, so is easily confused by the artificial lighting that we tend to use at the top and tail of the day.

* * *

Imagine a Scalextric car on a large oval track that takes 24.2 seconds to complete a loop. If I can see a stopwatch I can alter my speed so as to shave 0.2 seconds off my lap time. As I go round and round, I begin to get into a groove, squeezing and releasing the trigger at just the right time, until I hit upon the perfect rhythm that will carry me over the finishing line in 24 seconds precisely.

In this analogy, the timing of acceleration and deceleration is crucial. This is because a squeeze of the Scalextric trigger will affect the car's speed differently depending on whether it's going into or coming out of a bend. In the case of the SCN, the same dose of light delivered at different points in the molecular cycle can have completely opposite effects, the blue light of dawn normally speeding up the sequence of molecular events and the blue light of dusk typically slowing them down.

How is it that two people, one with a faster-than-normal and one with a slower-than-normal master clock, both manage to cross the molecular finishing line after exactly 24 hours? The answer is beguilingly simple. Everyone's SCN – from the preposterously fast to the laboriously slow – effectively writes its own photonic prescription. When the biological clock is working properly, it nudges us out of bed and drives us back into it in such a way that we get a highly individualised dose of light, one that's customised to perform the precise molecular correction that each of us needs to adjust the SCN to 24 hours precisely.

This helps explain why some of us are morning people, or 'larks', and some of us are evening people, or 'owls'. By waking early, the larks experience the blue light of dawn after an hour or two of wakefulness. The owls, by contrast, probably don't ever get to see the dawn and the blue light of dusk strikes much earlier in their biological day. Either way, the exposure to light is 'just so', exactly what's required to bring the molecular cycle into perfect line with the sun's 24-hour period.

The same phenomenon accounts for the two most common circadian disorders. If the SCN's molecular circuit happens to be exceptionally speedy, so comes full circle in significantly less than 24 hours, then your larkiness may be so extreme that you are said to suffer from advanced sleep phase disorder, or ASPD. This is characterised by waking in the small hours of the morning and going to bed in the early evening.

If the SCN ticks along particularly slowly, by contrast, then you will have the opposite problem. Extreme owls are diagnosed with delayed sleep phase disorder, or DSPD, the daily rhythm shifted so far forwards that you go to bed and get up much later than everyone else.

It is only relatively recently – in the 1980s – that the medical community formally recognised ASPD and DSPD as genuine phenomena, which means that up until that point those suffering from these conditions were largely ignored.

Emily Sloan, a woman who had suffered with what turned out to be DSPD since childhood in the 1920s, recalled being scolded, cajoled and threatened with punishment if, like a good little girl, she did not go to sleep at night. When in her 60s, she told researchers: 'I still feel guilty at my inability to fall asleep at the reasonable hour without tossing and turning for at least an hour.' In the 1930s, when Emily was aged about ten, her parents took her to the doctors. This ended up with her being prescribed sleeping pills, a bath and a walk around the block, the kind of bogus remedies with which narcoleptics are only too familiar.

Although the pills helped Emily get to sleep a little earlier, the struggle to wake up at the proper hour was immense. At college, she studiously avoided all morning classes. When, in due course, she had children and they were waking up at dawn, it was a massive strain. Her husband looked after the children before going to work, then turned them over to a live-in housekeeper. Emily slept until the late morning.

This regime put her at odds with the rest of the world. She avoided taking a regular job, fearing that she wouldn't be able to make it to work by 9 a.m. She made a point of organising her social activities in the afternoon and evening. In the 1960s, in her mid-40s, she was diagnosed with depression and went for psychoanalysis. It was only 20 years after that, aged 60, that she became one of the first to be diagnosed with DSPD, appearing as a case study in the paper that described the condition.

In recent years, research has started to reveal the genetic basis of ASPD and DSPD. Variations in genes like *Period*, *Clock*, *Timeless* and so on can result in radical changes to the SCN's molecular circuit, just as occurred in those mutated golden hamsters. People with ASPD and DSPD can synch to a 24-hour rhythm. They have to shift their wake and sleep times to do so but at least they can realise some kind of routine. In the case of another circadian disorder, routine is an unobtainable dream.

* * *

In non-24 sleep-wake disorder, or 'non-24', the process of entrainment fails completely. In the Scalextric analogy, this is like squeezing the trigger on the handset only to find that it's stuck. Without being able to accelerate and decelerate, it's impossible to correct the lap-time to the desired 24 seconds. This often happens for those who are completely blind. With no way for light to get to the hypothalamus, the SCN is unable to reflect the sun's daily rhythm to the rest of the body. Something similar can happen to sighted people too. Either way, the failure of entrainment means that all the cells round the body simply run to their own genetically determined pace, drifting out of tempo with the 24-hour regimen and with each other.

For the sake of simplicity, let's imagine a woman with non-24, whose unchecked SCN clock has a period of 25 hours. Like most of her friends, she sleeps for seven hours a night, but the point at which she feels sleepy slips an hour later every day. On day one, she goes to bed at 10 p.m. and wakes at 5 a.m. On day two, she can't get to sleep until 11. By day three, she is still awake at midnight. Come day 12, she is sleeping from 10 a.m. to 5 p.m. It is only after 24 days that she is back where she started.

In reality, most people with non-24 have a much harder time of things than this, because it's very unusual for the body's natural, genetically determined rhythm to overrun by as much as an hour a day. It would be much more common, for instance, to have a free-running cycle of 24 hours and 15 minutes, in which case someone with non-24 would only come into perfect alignment with the world around them once every 96 days, which is less than four times a year. And before they'd know it they'd be slipping out of phase again.

Even this does not capture the horror of non-24 properly. 'If you don't know you have it, you're going to try to fit in with what's going on around you,' says Marta Bravo, an artist living in Berlin, Germany. This inevitably means fighting to stay awake when your body is telling you to sleep and trying to sleep when you are at your most alert. Sometimes this works. Mostly it doesn't. All these efforts tend to mask the underlying pattern. 'It's not very often that you let yourself sleep when you naturally would, so it's hard to know what's happening.'

This makes self-diagnosis difficult, and like most other sleep disorders the awareness of non-24 is virtually non-existent. Marta remembers going to her GP and explaining how sleep was ruining her life. 'You're probably just depressed and you could try spraying lavender on your pillow.'

It took months to arrange an interview with Jonathan Patten. This was, in part, down to me being scatty and disorganised. But it was also pretty clear that the windows in which Jonathan was in shape for an interview were few and far between. When he suggested a date to me via Twitter and I missed his message, he went to ground for several weeks, finally resurfacing again to offer another opportunity to talk.

At the age of 13, Jonathan was diagnosed with Tourette's syndrome, attention deficit hyperactivity disorder, obsessive-compulsive disorder and anxiety. But the worst of his problems, by his own admission, is his wildly variable circadian rhythm.

'It started off with these bizarre sleeping schedules,' he says, remembering his time as a student at Covenant College in Georgia in 2010. He tells me about one week in particular, where his body clock seemed to be completely opposed to the real world. 'I was awake at four in the morning, at five, at six,' he says. At seven o'clock, he finally passed into sleep only to wake up later in the day having missed all his classes. This resulted in his suspension from college for poor attendance and bad grades. Only then did he begin to confront the possibility that he had a real problem.

In parallel with his academic struggle, Jonathan's social life began to suffer. If he'd arranged to meet someone at a particular time and then didn't show, they'd think he'd let them down deliberately. 'At college, I lost 95 per cent of my relationships to my sleep disorder,' he says. 'It sucks.' Jonathan still worries about how others will interpret his behaviour. 'I want to let them know that I'm struggling with a sleep disorder, but will they judge me, will they care, will they label me as crazy?' Questions like these are difficult to suppress. For many people with disordered sleep, the solution is simply to go to ground and avoid commitment altogether.

In short, planning is a near-impossible task for someone with non-24, and the inability to anticipate how you will be feeling tomorrow, next week or next month can drive those afflicted into an isolated present.

\* \* \*

Apart from APSD, DPSD and non-24, there are plenty of other ways in which the body can get out of synch with the sun.

Every spring, many countries around the world reset their clocks so that they are in daylight saving time. This spring forward by one hour steals roughly 40 minutes of sleep on the Saturday/Sunday night of the clock change. The resulting sleep deprivation may explain why the following Monday is often referred to as a 'sleepy Monday', a day on which there is a demonstrable spike in accidents in the workplace and on the roads. There are even data to suggest that judges tend to offer harsher sentences than normal on the Monday following a spring forward to daylight saving.

Boarding an aeroplane and chasing the sun to the west or fleeing from it to the east poses a similar biological challenge, one that is proportional to the number of time zones crossed. As I suffer from narcolepsy, I've never been unduly bothered by jetlag, its most obvious manifestation of needing to sleep at strange times being something that I'm fairly used to.

On a flight from London to California, however, I pay special attention to the mixed messages my brain receives as I travel across eight time zones in eleven hours, and I do my best to drive my brain as hard as possible towards the local time at my destination.

In the departure lounge before my lunchtime flight, there is the inevitable temptation to have a bite to eat, but I resist. This would be unwise because the surge of carbs, protein and fat following a meal act as another 'zeitgeber', an important time cue that also has an influence on the SCN rhythm. I delay this message, eating lunch several hours later when I'm in the air.

As the flight wears on and it's getting dark at home, I figure I need as much natural light as possible. On the port side of the aeroplane, the side facing the sun, most of the windows are blacked out or manually set to a blue tint. There is one passenger, however, who has not dimmed his window and as luck would have it I can position myself so that I am staring across the cabin and straight into the sun. To starboard, on my side of the plane, I look down at an unbelievable expanse of sea ice cracking its way to the curved, hazed horizon at the top of the earth. At 38,000 feet, I haven't got a hope of spotting a polar bear, but I stare at the blue-and-white landscape anyway, hoping my brain takes the hint.

By the time I've landed in California, it's 10 p.m. local time and I've been awake (more or less) for 22 hours. Before I fall asleep, I notice that I'm shivering violently. In spite of my efforts to shift my body clock towards Pacific Standard Time, the chills suggest I haven't been particularly successful. My SCN is clearly still operating on Greenwich Mean Time, where it's 6 a.m. and my circadian cycle and hence body temperature is at its nadir. In reality it takes several days for the brain and body to cope with such radical time travel.

Night shiftwork often results in the same physiological chaos. Even after working for five consecutive nights and sleeping through five consecutive days, the SCN circuit will not be well adjusted. Come the weekend, shiftworkers tend to revert to operating in the hours of daylight. With mixed light messages pulling the SCN in different directions, the gut, the liver, the kidneys and the immune system, like so many metronomes, drift out of synch. This may explain why night shiftworkers are at greater risk of cancer, obesity, cardiovascular disease, type 2 diabetes and a host of gut problems. Even with blackout blinds, an eyepatch and earplugs, sleep in the hours of daylight is usually more disturbed than sleep during the hours of darkness. So shiftworkers will often be in a constant state of mild deprivation too.

* * *

Based on the powerful role the sun plays in the entrainment of the biological clock, it's not surprising that the changing day length that occurs with the seasons can result in circadian problems. Indeed, the extremes of light and dark seen at the poles have a clear effect upon the SCN rhythm, patterns of sleep and both mental and physical health.

The psychological impact of living in constant darkness was obvious to early polar explorers. On board the *Belgica*, the first ship to winter in the Antarctic in 1898/1899, there was widespread disturbance amongst the Belgian crew. The ship's surgeon Frederick Cook noted that 'the outlook was that of a madhouse'.

When the sun eventually returned, there was always much rejoicing, as occurred in 1915 after Ernest Shackleton's crew of 28 had spent 92 consecutive days in the dark waiting for the pack ice to free up their ship, the *Endurance*. First engineer Lewis Rickinson recorded that the day the sun returned was 'treated as Christmas Day', with 'the special feast of duck and green peas, fresh – out of a tin'.

In the 1930s a German woman called Christiane Ritter travelled to the polar north to be with her fur-trapper husband and made beautiful, insightful observations of the light and landscape, particularly around the onset of the polar winter. 'It is at precisely this time that a decisive change takes place in the human mood, when the reality of the phenomenal world dissolves, when men slowly lose all sense of fixed points, of impulses from the external world,' she wrote in *A Woman in the Polar Night*. 'Those who are accustomed to yield to their inclination to idleness run the great danger of losing themselves in nothingness, of surrendering their senses to all the insane fantasies of overstretched nerves.' This, she felt, could explain a phenomenon familiar to those dwelling at extreme latitudes: 'polar mentality'.

But it isn't even necessary to go to the poles to experience how differing day length can affect biology. In 1913, German psychiatrist Emil Kraepelin noted that some of his patients, with what would now be called bipolar disorder, began to show signs of moodiness in autumn, a behavioural change that lifted in spring, 'when the sap runs in the trees'. These emotional changes, he argued, were even detectable in 'healthy individuals at the changes of the seasons'.

It was not until the 1980s, however, that this phenomenon began to be more widely recognised. In 1981, with a bunch of circumstantial evidence that mood changed with the seasons, researchers at the National Institutes of Health in Bethesda, Maryland, encouraged the *Washington Post*'s health correspondent to run a feature about one of their patients.

Ingrid Bush (as the journalist referred to the patient) suffered from depression every winter and mild mania in the spring. These patterns, she noted, seemed to be exacerbated by latitude, so the further north she was living the worse the depression would be and the longer it would last. Interestingly, there were two winters when she'd taken herself to Jamaica and on both occasions she'd floated out of her depression within two days of touching down. At the end of the article, the researchers smuggled an invitation for anyone else experiencing seasonal mood changes to get in touch. They received more than 2000 replies.

With further screening and some very strict criteria, the researchers diagnosed around 30 of these potential patients with 'seasonal affective disorder' or SAD. In addition to describing feelings of sadness, anxiety and irritability during the winter months, many reported craving carbohydrates and putting on weight. All but one of the subjects said they slept longer during winter nights and found themselves drowsy and unrested during winter days.

Thomas Wehr, the senior author on that seminal SAD paper, went on to conduct several neat experiments showing that changing day length can affect circadian biology and, hence, sleep. But there's one in particular that is of special interest. He invited volunteers into his sleep lab, a windowless world cut off from all obvious circadian signals, and one in which Wehr used artificial lighting to manipulate day length. He began by settling his subjects into the kind of diurnal cycle typical of a summer's day in Maryland, with 16 hours of light and eight hours of darkness. Under these conditions, the volunteers experienced textbook sleep, settling quickly and remaining unconscious until the lights came back on in the morning.

Then, after a week or so on this schedule, Wehr switched them onto a rather more wintry scenario, with short, ten-hour days and long, 14-hour nights. This threw their sleep into disarray. What was really surprising, however, was that after a couple of weeks on this schedule, the volunteers' sleep began to fracture into two clearly distinguishable chunks, one in the early evening followed by several hours of wakefulness and then another sleep.

This was peculiar because most people don't notice much of a change in sleep from one season to the next. Wehr had an idea why. 'In the past 200 years, humans have developed increasingly efficient lamps and inexpensive sources of energy to power them,' he wrote. There has been a migration from the countryside to the cities too, from a life outdoors to one indoors. 'Consequently, humans have increasingly insulated themselves from the natural cycles of light and darkness that have shaped the endogenous rhythms of life on this planet for billions of years.' Thanks to artificial lighting, the human brain has become 'perpetually clamped in a long-day/short-night mode,' he wrote. Implicit in Wehr's thesis was the prediction that in the past, when winter nights were long and dark, we might have slept in two discrete stints.

When social historian Roger Ekirch chanced upon a

write-up of Wehr's work in the *New York Times* in 1995, he was stunned. He was researching sleep as part of a book on the history of night-time, and though he had no prior knowledge of Wehr's research he'd arrived at much the same conclusion.

In the mid-1980s, Ekirch had spent a portion of his summer holidays in London, delving into the archives at the Public Record Office on Chancery Lane. He knew from his previous work on the banishment of convicts to the US in the early eighteenth century that legal records were a goldmine for historians like himself. As a lot of criminal activity takes place under cover of darkness, they were a particularly valuable resource for his work on the night.

As Ekirch worked his way through these nefarious stories of thieving and skullduggery, he began to notice references to 'first sleep'. The tragic deposition of a nine-year-old girl called Jane Rowth had a particular impact.

One dark night in 1697, Jane's mother had woken 'after shee had got her first sleep' and 'gotten out of bedd'. Jane too was awake, it seems, because she told the court that her mother had been 'smoaking a pipe at the fire side' when two male companions appeared at a window. They 'bad her make ready & come away', presumably to carry out some preplanned felony. Before she left, Jane's mother instructed her daughter to 'lye still, and she would come againe in the morning'. She never did. Her dead body turned up a couple of days later.

Ekirch was intrigued by the girl's use of the term 'first sleep'. By the time he encountered Wehr's work a decade later, he had come across dozens of similar references to first sleep, second sleep and a clear interval of wakefulness in between.

In many of these historical accounts, the first and second sleeps were clearly practiced throughout the year and not limited to the winter months. But it's easy to imagine that before the introduction of artificial lighting, the biphasic pattern of sleep might have been most pronounced during the interminable darkness of a long winter night in the northern hemisphere. 'How long the nights are,' wrote American novelist Nathaniel Hawthorne in 1854, 'from the first gathering gloom of twilight, when the grate in my office begins to grow ruddier, all through dinnertime, and the putting to bed of the children, and the lengthened evening, with its books or its drowsiness, – our own getting to bed, the brief awakenings through the many dark hours, and then the creeping onward of morning. It seems an age between light and light.'

With such extended periods of darkness, it made sense to do something profitable. 'Families rose from their beds to urinate, smoke tobacco, and even visit close neighbors. Remaining abed, many persons also made love, prayed, and, most important, reflected on the dreams that typically preceded waking from their "first sleep",' wrote Ekirch in an article he published in 2001. 'Until the early modern era, Western Europeans on most evenings experienced two major intervals of sleep bridged by up to an hour or more of quiet wakefulness.'

In the 35 or more years since Ekirch first noted descriptions of segmented sleep, he has amassed thousands of similar references. He has uncovered mentions of 'first sleep' and 'second sleep' in no fewer than 15 different languages. The earliest appearance he's come across appears in Homer's *Odyssey*, when Proteus's daughter Eidothea tells Menelaus to seize her father in 'primo somno'; the Italians have 'primo sonno'; the French speak of 'premier sommeil'.

In his late 20s, Robert Louis Stevenson captured this sense of middle-of-the-night wakefulness rather beautifully during a trek he made through southern France in 1878, accompanied only by a donkey called Modestine:

There is one stirring hour, unknown to those who dwell in houses, when a wakeful influence goes abroad over the sleeping hemisphere, and all the outdoor world are on their feet. It is then that the cock first crows, not this time to announce the dawn, but like a cheerful watchman speeding the course of the night. Cattle awake on the meadows; sheep break their fast on the dewy hillsides, and change to a new lair among the ferns; and houseless men, who have lain down with the fowls, open their dim eyes and behold the beauty of the night.

By this time, the biphasic pattern of human sleep had already vanished from much of the world, banished from our behavioural repertoire and collective consciousness by the invention of electric lighting as Wehr envisaged. This was a protracted evolution, says Ekirch, a transition that took place over the course of the nineteenth century, the interval of wakefulness gradually being pushed further and further into the night until it became reduced to little more than a brief doze just before getting out of bed.

The combined case made by Wehr and Ekirch's research is compelling, and underscores the tremendous impact that light can have on when and how we sleep.

* * *

Beyond ASPD, DSPD and non-24, jetlag, shiftwork and latitude, it's also clear that the sun provides us with a different kick at different times in our lives.

As any parent will know, the first few weeks of a baby's postnatal life are chaos, with waking, sleeping and feeding at odds with the circadian expectations of parents. By about three months, however, most babies will be beginning to show an appreciation that they exist in a world where the sun is of fundamental significance; there is more wakefulness during the day and more sleep at night. It's the most they will ever sleep, roughly between 14 and 17 hours a day in a 24-hour period.

When I became a father, I spent many nights cradling my newborn son in the darkness, hoping that perhaps the next night would be the one that he managed to achieve a complete division between the day and the night and sleep through the small hours without screaming. He had passed his first birthday before he pulled off this feat. There was much rejoicing. At the age of 13, my son is now about to enter puberty and I am preparing myself for another shift in his body clock.

Teenagers find it harder and harder to get out of bed in time for school and want to stay up later and later at night. Come the weekend, their lie-ins are legendary. This relaxed attitude is thought to explain the twinned phenomena of 'Sunday night insomnia' and 'Monday morning blues', something that many adults experience too. Having spent much of Sunday morning in bed, it's hard to get to sleep that night, resulting in a state of mild sleep deprivation on the Monday morning.

Parents of teenagers often interpret such behaviour as laziness, bursting into bedrooms, dragging off duvets and unceremoniously turfing their children out of bed so that they'll make it to school on time. In the 1990s, however, evidence began to emerge that at least some of this apparent slothfulness can be explained by a real, biological change. Sleep researcher Mary Carskadon and colleagues put a questionnaire to a bunch of American sixth graders (typically 12 years old).

'Imagine: School is cancelled!' began the first of ten questions. 'You can get up whenever you want to. When would you get out of bed? Between:

a. 5:00 and 6:30 a.m.

b. 6:30 and 7:45 a.m.

c. 7:45 and 9:45 a.m.

d. 9:45 and 11:00 a.m.

e. 11:00 a.m. and noon'

In addition to posing questions like these, Carskadon collected other data on the children, including measures of their social and physical development. Their findings, tentative at that stage, suggested that the onset of puberty might be causing a real shift in the body clock.

In just about every country and culture that researchers have subsequently looked at, the story of teenagers and sleep is remarkably similar. As they age, the onset of melatonin release is progressively delayed, and they stay up later and later at night. At weekends, without the need to get to school, they tend to lie in for longer.

Interestingly, there appear to be similar changes taking place in juvenile animals. During what would be its teenage years, for instance, the lab rat experiences a shift in its body clock of at least an hour. Likewise the rhesus monkey, the degu and the delightfully named fat sand rat.

The factors that play into this shift in the teenage body clock are undoubtedly complex. As children mature, they have more demands on their time, like more homework and a pressing need to socialise. In their natural ambition to become fully fledged adults, it makes sense that they should try to stay up at least as late as their parents.

This is tiring in itself, but there's now good evidence that in a teenage brain the molecular clock becomes especially sensitive to the light at the end of the day, and this may exacerbate the problem. If a teen spends the extra hours of wakefulness staring at a tablet or mobile, for instance, it's likely to have a particularly strong effect, pushing the onset of sleep until the early hours of the morning.

The obvious fix would seem to be to give teenagers a break, letting their body clocks beat to the new, later rhythm, and simply delaying school start time by an hour or two. Plenty of schools have tried this, reporting that under the new regime their students get more sleep, are more likely to be on time, are less moody, more motivated and get better grades. In the long-term, however, it looks as though children accommodate to the new start times, simply staying up even later than they would have done, ending up with just the same amount of sleep as before and being just as tired during the first few lessons.

Once the physiological stress of puberty is over, the body clock tends to settle back down to a stable rhythm. But as we get older, the clock seems to change again, shifting us to ever earlier wake-up times, so that a common complaint in the over 60s is waking in the early hours of the morning. The way in which the body clock shifts with age may just be an unavoidable consequence of some biological exigency we have not yet worked out. An alternative and rather ingenious possibility is that these differing sleep patterns constitute an evolved shift-work system.

Step back a hundred thousand years or so and imagine your immediate ancestors, an extended family of troglodytes with their hunter-gatherer struggle for survival. If everyone in the clan – from newborns to the elderly – had perfectly synchronised biological clocks there would be huge swathes of the night where the fire was left unattended with no one keeping watch for wolves. By contrast, a family in which teenagers are programmed to stay up into the small hours and grandparents pick up the slack a few hours later might be at a selective advantage. 'It provides coverage for the night,' says Zeitzer. It's impossible to test, he admits, but it's a fascinating notion nonetheless and one for which there is at least some circumstantial evidence. A study of the Hazda, a hunter-gatherer society from northern Tanzania, found that over the course of 20 days of observation of around 30 individuals there was just 18 minutes when everyone was asleep.

Given the many ways in which our internal biology can stray from the predictable rhythm of the sun, all the biological chaos that ensues and the negative consequences for physical and mental health and sleep, it would be rather nice to have more control over the SCN's orchestrating rhythm. If there were a way to tinker with the internal biological clock, it might be possible to correct DSPD or ASPD, to manage jetlag, shiftwork and seasonal affective disorder or to pull back the teenage clock by an hour or two.

With the evidence that light and melatonin have a key role in keeping the circadian rhythm to time, so an industry has emerged marketing light boxes and melatonin tablets, suggesting that even those without sleep disorders stand to benefit from more light and supplementary melatonin. If there is no underlying circadian disorder, however, such interventions are unlikely to improve sleep, says Zeitzer. At best these measures will be costly but largely ineffective, bringing about negligible improvements to sleep that could easily be attributed to the placebo effect. At worst, artificial light and melatonin tablets, if delivered in the wrong dose or at the wrong time of day, could actually upset the SCN's careful orchestration of the body and, hence, sleep.

That said, there's considerable promise in going beyond the basic light box to get much, much smarter about the way we interact with light.

\* \* \*

Zeitzer's sleep lab looks a little like a Tracy Emin installation: the unmade bed with rumpled sheets in the middle of the windowless room; a side table littered with DVDs; a rack of miscellaneous electrical equipment and wires; an ophthalmologist's chart; a trolley strewn with balls of tissue, a pair of used latex gloves and a huge lamp directed at the empty bed. But it's not a work of modern art, it's part of Zeitzer's ongoing research into the impacts that light at night can have on the biological clock.

In this particular experiment, the lab technician wakes up a sleeping subject in the middle of the night, switches on the lamp and chats to them for an hour or so. On one night, the volunteer is exposed to a standard bulb emitting the complete spectrum of visible wavelengths, as would occur in most homes today. On another visit, the bulb is one devoid of blueness, which should result in less circadian chaos. By monitoring electrical signals on the scalp and through cognitive tests, Zeitzer is looking to see if the different light experiences bring about differences in alertness. 'We also see how long it takes them to fall asleep afterwards. If the light is really waking you up, it's going to take you longer to fall asleep.'

On the way back to his office, Zeitzer anticipates the implications of this kind of research. 'There's lots that we could and should be doing to improve our circadian health,' he says. The challenge is to convince people that their relationship with light really matters, not just for their sleep but for their health too.

'Smart lighting' could help in this regard. There is already a handful of hotels toying with such technology, manipulating the wavelengths of light in rooms in an effort to help jet-setting guests adjust to their new time zone. 'If you are just in town for a day, a few hotels will actually keep you in your home time zone if you want,' says Zeitzer. Some hospitals are doing something similar, transforming intensive care units from a 24/7, fully fluorescing environment into one where strip lights deliver a far more natural sequence of wavelengths.

In the not-too-distant future, it seems inevitable that our homes will be rigged up to deliver a personalised blend of wavelengths to maximise our performance too. In my home, for instance, there might be a straight-forward prescription of bright, white light topped and tailed by 15 minutes of intense blue at dawn and dusk. The bathroom – often the brightest room in the house in which many of us spend several minutes just before trying to sleep – is an obvious place that would benefit from more sensitive lighting. In the middle of the night, the lights would probably be less intense and more reddish in colour, just enough for us to navigate safely to the bathroom and back but not enough to interfere with the circadian clock.

Before Zeitzer's tour concludes, he leads me into the office next to his own and introduces me to Manuel Spitschan, a postdoc with whom he has been pursuing another fascinating line of circadian enquiry that could drag the brains of teenagers closer to the social world.

This work stems from a chance discovery made by another Stanford researcher in 1998. Craig Heller and colleagues found that exposing mice to a series of brief flashes of bright light could delay or advance their body clocks by several hours depending on the time of exposure. Others found that something similar happened in rats. Another group showed the same effect in hamsters.

Armed with a dossier of patents, Heller managed to persuade

Zeitzer to collaborate on an investigation into humans. 'I thought this had no chance of working,' admits Zeitzer. But it did. They used a similar protocol to the one Heller had used on mice, exposing volunteers to a brief pulse of light once a minute for an hour during the night. In comparison to control subjects, who spent the hour in the pitch black, flashing lights delayed the biological clock by almost an hour. The implications are obvious. Flashing lights might be a much more effective (not to mention energy-efficient) way of treating circadian disorders.

Spitschan is fondling what looks like a pair of ski goggles made of red plastic, except that instead of a see-through lens, there's an opaque screen. Incredibly, pulses of light delivered to this eye mask can influence the circadian clock even during sleep. 'About 10 per cent of light gets through the eyelids,' says Zeitzer.

I get myself comfortable in a chair and pull the goggles on over my eyes. Spitschan fiddles with some settings on his computer so that I get a 10-millisecond flash of light every 15 seconds. I close my eyes, as if asleep, and he hits the start button. The pulse, when it comes, bleaches my retina. It would definitely keep me awake and if I were asleep it might even wake me up. Spitschan rejigs things so that I receive a pulse that is 1000 times shorter. I am aware of it, but it's so brief that I would probably be able to kip off and it would be unlikely to disturb my sleep.

So far, Zeitzer has been able to realise the biggest shift to the body clock by setting the light to flash at eight-second intervals. 'We are still exploring why this is the case, but it seems to be corroborated by independent evidence from rodents,' he says. 'There's sufficient data to say that it works.' Teenagers are just the start of it. 'The question we are asking is can we optimise it further,' he says. 'Instead of getting a couple of hours' shift, can we get 5, 6, 7, 8, 9, 10 hours?' If that were possible, the range of applications would be tremendous.

\* \* \*

Back in the UK, I travel to Guildford to interview Zeitzer's sometime collaborator Derk-Jan Dijk, a circadian biologist at the University of Surrey. He is very clear that light is an issue of public health. 'We are concerned about the quality of the air we breathe and put certain measures in place to make sure that it's not becoming too unhealthy,' he says. 'We are well aware of the effects of caffeine consumption late in the evening on subsequent sleep. We just need to start thinking about light in a similar way.'

Figuring out the many mechanisms through which light affects us is still work in progress, but there is compelling evidence that it does. When researchers look at how gene expression changes over a 24-hour period, there are clear and highly regular patterns. Disturbances to the circadian rhythm cause huge disruption to these patterns of expression.

As individuals, the simple awareness that our exposure to light is affecting our biology and health is crucial. Dijk makes two specific and related recommendations: 'Most of us are now working inside during the daytime and get far less daylight than we used to,' he says. Finding ways to increase the exposure to light during the day – either natural or artificial – will reduce our sensitivity to the messy, highly variable exposure to light in the evening. It is also important to avoid too much exposure to artificial light in the early evening, particularly for those who've been stuck indoors all day.

At the societal level, there are also beneficial measures that could be taken. Light pollution at night is a big issue. In 2016, researchers got their hands on satellite images from across the US at night, ones that revealed how levels of light pollution differed from one street to the next. Then they picked up the phone and called around, conducting interviews with residents to get an idea of their sleep patterns. Living in an area with high light pollution was associated with an increased likelihood of sleeping less than six hours per night. More light was linked to a clear shift in the average bed time and wake-up time. Brighter environments were also correlated with the risk of sleeping badly at night and sleeping more during the day. This and other studies like it fall short of demonstrating that light at night caused these changes in sleep, but from all that we know about the effect of light on the circadian system it more than likely played a role.

Where Dijk lives, his council has taken the decision to switch off street lighting at night, partly to save on electricity but also because of this kind of disruptive impact on sleep. When designing the homes and workplaces of the future, there needs to be far more attention paid to the importance of light, he says. At work, employers can also make simple changes to lighting systems that could improve the mental health and hence the productivity of employees. There's compelling evidence that exposure to light-emitting devices like mobile phones or tablets late at night is misleading our brains and bodies. Simple tweaks to software, such as Apple's Night Shift, which steers the wavelength of emitted light towards the red, less disruptive end of the spectrum, are a good start.

\* \* \*

With a clearer understanding of how the circadian rhythm is established and what affects it, let's turn the spotlight back to narcolepsy to take a closer look at one of its most extraordinary symptoms.

# Weak with laughter

*'Mirth is like a flash of lightning, that breaks through
a gloom of clouds, and glitters for a moment.'*
Joseph Addison

I was at a conference, standing in the queue for coffee during a break between sessions, and the woman in front of me went down. As she fell, she resembled a push puppet, one of those little elasticated toys that collapses when you press the button on the base. It all happened very quickly, but if it had been possible to slow down the motion, I would have seen her head drop first, chin onto chest, her shoulders relax, arms flop to her sides, and legs buckle.

Within seconds, and before I could offer my assistance, she was back, rising Phoenix-like inside her bright orange dress in perfect time to receive a cup of coffee and a biscuit. I would have been surprised at this fit, except that this was the annual meeting of Narcolepsy UK, and falling over like this is standard practice for most people with narcolepsy. To the uninitiated, it might have looked as though the woman in front of me had suddenly fallen asleep. I knew better. In spite of appearances, the woman in the orange dress had remained completely conscious throughout this brief episode.

This is cataplexy, a condition in which emotions can cause the body's muscles to fail; it affects many people with narcolepsy.

Nathaniel Kleitman understood the difference between narcolepsy (the sleep) and cataplexy (the collapsing fits) only too
well. 'Boredom and monotony favor narcolepsy; gaiety and
excitement, cataplexy,' he wrote in *Sleep and Wakefulness*.

For those who have first-hand experience of narcolepsy *and*
cataplexy, the distinction is abundantly clear. Most obviously,
narcolepsy (that invincible need to sleep) involves a loss of
consciousness. Cataplexy does not. This is underscored by
the fact that cataplexy has an extraordinary number of ideophones, invented words that offer powerful testimony of the
fact that this is a conscious state that can be perceived.

There are a few people whose words for cataplexy were
probably coined in the first few weeks, when mirth resulted
in a new and puzzling weakness at the backs of the knees but
before this progressed to a full-blown collapse. In her brilliant
memoir *Wide Awake and Dreaming*, Julie Flygare refers to this
feeling as 'my knee thing'. Others have jelly legs, heavy legs,
spaghetti legs, a funny turn, a moment, a melting moment.
More commonly, there are simple nouns, distilling each attack
into a clearly defined, almost sentimental entity. Where I have
'a gibber', for instance, others have a wobble, the wobbles, the
wobblies, the jitters, a pyjama flop, the floppies, the jellies, a
jellyfish attack, a cat attack, a cat do, a do. Many use verbs
that give a better sense of the action. If I'm in the middle of
an attack, for example, I am gibbering. Others swoon, bungle,
fall from grace, go lop sop doi. More rarely, there are simple
exclamations, like 'oh no' or 'man down', succinct utterings
that alert those in the know to an impending collapse. Doctors
too have suggested plenty of names for cataplexy over the
years, like affectotonia, gelolepsy, affective adynamia, tonus
blockade, emotional asthenia and geloplegia.

Giuseppe Plazzi, head of the sleep lab at the University of

Bologna, has argued that Dante Alighieri might have suffered from narcolepsy with cataplexy all the way back in the fourteenth century, as his autobiographical masterpiece *The Divine Comedy* features most of the symptoms, including cataplexy. In the middle of his journey through Hell, for instance, Dante hears the tragic love story of two lost spirits and collapses. 'I fainted out of pity, and, as if I were dying, fell, as a dead body falls.'

The idea that Dante suffered from narcolepsy is certainly intriguing, but most sleep specialists – including Plazzi – date the first unequivocal description of cataplexy to 1877, when German psychiatrist Karl Westphal presented a case at a meeting of the Berlin Medical and Psychological Society. He had witnessed several attacks in one of his patients, a bookbinder called Herr Ehlert. 'His eyes close involuntarily, and he cannot keep them open,' Westphal told his medical audience. 'At the same time, he loses all strength in his limbs and the ability to speak. He cannot move, and must sit or lean on something … He reports that he hears and understands what is said to him during an attack.'

On another occasion, Westphal watched Ehlert 'staggering like an intoxicated person', then standing and swaying but without falling. 'During this time, slight twitching movements in the face were observed, as were movements of the jaw.' To Westphal, it looked as though Ehlert was searching for a chair or seat to hold himself up. When he finally found a support, he murmured 'Chair', and with eyes half-shut added, 'Professor, please excuse me while I take a seat.'

When Gélineau came to describe the narcolepsy of the cooper Monsieur 'G' a few years later, he also described clear-cut cataplexy. 'When laughing out loud, he would feel weakness in his legs, which would buckle under him,' wrote Gélineau, but he went on to equate these attacks with sleep.

'If he experiences a deep emotion … the need to sleep is even more urgent and sudden. Thus, for example, if he is closing a good business deal, if he sees a friend, if he speaks with a stranger for the first time, or if he receives a good hand while playing cards, he collapses and falls asleep,' he wrote, failing to make a clear distinction between his patient's sleepiness and his collapsing in response to emotions.

This is a common mistake, even today, most people failing to distinguish between the sleepiness of narcolepsy and the cataplectic seizures that have an outward resemblance to sleep but are, as we will see, something completely different. Whilst it's great that more and more people have heard of narcolepsy, it's a shame that the cataplexy that many of us experience is not more widely known, not least because it's so interesting.

* * *

For many people with narcolepsy, the excessive sleepiness is the first symptom. The cataplexy, if it comes, usually takes weeks, months or sometimes years to develop. It is most commonly triggered by a joke. But the context is crucial.

If I am alone, watching TV perhaps, and a comedian delivers a witty one-liner, I might smile or even chuckle but I will never, ever have cataplexy. In company, however, that exact same joke could bring on a full-blown fit. The identity of my companions matters too. If we are not well acquainted, I am far less likely to gibber than if I am cosseted by close friends and family. The source of the trigger is important too. Rather embarrassingly, it's my own cheeky thoughts or jokes that are most likely to cause cataplexy. If a friend makes a gag, I might chortle but no more. If I made the same joke and it caused a ripple, I'd be down on the ground in no time.

This tallies rather nicely with what other people with cataplexy report. In a recent survey, the researchers recognised laughing as the most common trigger, but correctly noted that 'other qualities need to be associated with laughter to reliably evoke cataplexy'. Laughing excitedly was the most frequent trigger. Telling a joke was more likely to cause a collapse than hearing a joke and some 85 per cent of respondents were familiar with falling over before reaching the punchline. The laughter resulting from being tickled was more effective than the laughter as tickler. One of the most reliable triggers was making a sharp-minded remark.

It was a sharp-minded remark that caused me to experience my first major cataplectic attack back in 1994. I was back from college for the Easter break and was over at the home of my best friend Zaid, watching television in his bedroom. We were lounging on beanbags on the floor. The weather forecast came on. The presenter had somewhat exaggerated incisors. I let out a gentle neigh. Zaid snorted. It was juvenile in the extreme. It was unkind. If not quite sharp-minded, it was certainly barbed, and in that moment, shared between close friends, I slumped back into the beanbag in a fit of uncontrollable hysterics.

The mirth rippled through my body, but as my muscles shuddered, then failed completely, the laughter didn't last long. It felt as though my breathing had stopped and my heart was faltering. I was sure I was dying. The thrill turned quickly to fathomless dread. I panicked and tried to call out, but I could not move my lips and no sound came, leaving the distress to echo in silence inside my head. I heard my name. Zaid had noticed that something was wrong. 'Henry?' he repeated, this time with greater urgency. Then he reached across and shook me, his grip bringing my near-dead body back to the world of the living. I sat up and burst into a cascade of tears.

Sarah Garvey has a similar story of how a shared moment, even if it's not particularly funny, is enough to trigger an attack. She became aware of sleepiness when she was just 14, then began to develop cataplexy and eventually got a diagnosis of narcolepsy with cataplexy when she was 16. When I first spoke to her in 2013, her mother had rarely witnessed her cataplexy, and she had yet to have an attack in public. But her two sisters – one older and one younger – soon learned how to take their sister down. 'The best person to bring on my cataplexy is my older sister,' says Sarah. 'She knows how to make me laugh. She has showed me the stupidest things.'

On one occasion, Sarah's sister pulled up a picture on the internet of a college student dressed up for a Halloween party. Rather than going as a conventional ghost, vampire or zombie, he'd combined two characters from children's fiction: Dumbledore, the sage headmaster at Hogwarts School of Witchcraft and Wizardry and Dora the Explorer, a cartoon character for the under fives. Dumbledora the Explorer sports the wizard's wand, spectacles and long-grey beard and the young adventurer's trademark orange shorts, pink T-shirt, purple backpack and bobbed black hairdo. It's silly enough, I can see, to trigger an attack.

\* \* \*

The first time this happens, it's alarming. But the recovery is so quick that it's easy to dismiss as a one-off. This is what I did with that first attack in Zaid's bedroom. Over the next few weeks and months I had more episodes, but they were relatively mild and fleeting, so when I eventually went to my doctor, in the summer of 1994, I only reported occasional weakness in my legs when I laughed. As I didn't seem too alarmed, he wasn't either. It was unlikely to be serious and would probably pass, he told me. I biked back to my college, relieved.

By the end of the summer term, however, the attacks had become more frequent and stronger, though as with the sleepiness, the change had been so incremental that neither I nor my friends noticed the extent of the deterioration. So it was only back at home that I took a moment to reflect on the extent of my transformation.

I was with my brother and sister, neither of whom I'd seen for months, and when I toppled to the kitchen floor in a heap of mirth, they were extremely alarmed. I was about to go to India over the summer and, with my parents away, my siblings insisted I see a neurologist before I went. After some phone calls to my GP and a family friend, I managed to get an appointment with a consultant neurologist at the National Hospital for Neurology and Neurosurgery in Queen Square, London.

I took a seat in the marbled entrance. There was a man next to me, his limbs jerking with alarming frequency. A woman shuffled to the bathroom, her head ticking back and forth like a chicken. 'Fuck it,' shouted a youth coming down the staircase. I was in the wrong place, surely?

The consultant was a tall, thin, serious man, grey and quiet. He listened as I told him about my peculiar symptoms, making occasional jottings in what were to become my medical notes. I explained that I kept falling asleep in lectures and in the library, which didn't seem to surprise him. 'We all do,' he said, or something to that effect.

I went on to describe how certain types of joke would provoke uncontrollable laughter, how it felt like I couldn't breathe, how my lips, my eyelids and my limbs would quiver, how I'd feel weak and might drop things. 'Knows before the attack, that it's going to happen. Friends say he looks very odd – as if paralysed during episodes,' he wrote. 'Tries to speak, but unable to,' he recorded, but 'no loss of consciousness'. The consultant wondered whether I might be suffering from 'a type of reflex seizure', a 'very rare form of epileptic attack'.

This experience is strikingly similar to that of a 14-year-old girl called Olivia, who had entered the exact same hospital some 70 years before me with the exact same symptoms and received the exact same misdiagnosis. While Olivia's neurologist sent her packing with a prescription of anti-epileptic drugs, mine sent me for brain scans. 'I doubt that there is any abnormal neurological basis for these attacks and I see no reason to limit his activities at the moment,' he wrote to my GP. I went to India.

\* \* \*

Although wise-cracks are one of the best ways to bring on a cataplectic attack, there are others. It was at the end of my final year at university and more than one year after I'd shown the first signs of excessive sleepiness that I had my first taste of a non-humorous trigger. 11 June 1995: it was the Rugby World Cup and England was up against Australia in the quarter finals. Around 50 undergraduates had crammed into the common room, beers in hand, to watch the match, and at full-time it reached an impossibly exciting climax, with the rivals tied at 22 apiece.

Then, deep into injury time, the ball came back to the English fly-half Rob Andrew, and at 45 metres away from Australia's goal-line, he unleashed a mighty drop-kick. The ball began to rise. All eyes – on the pitch, in the stadium, in front of every TV broadcasting the match – followed the oval as it sailed in what appeared to be slow-motion through the uprights. England had won a magnificent 25–22 victory. 'People jumped 10 feet when that went over,' said England coach Jack Rowell at the time. 'When they came down to earth, they were crying.' This is pretty much what happened in the common room, except that as I rose I fell. Everyone around me was jumping up and down screaming, hugging each other, beer flying, so engrossed in the moment that they didn't notice me slumped, almost horizontally in my chair, eyes closed, my lips twitching into a flickering grin. I'd been floored by elation.

The following year I discovered that cataplexy can also be brought on by surprise. As a recent zoology graduate, I'd travelled to the Kalahari Desert in South Africa to study meerkat behaviour, and it was here that I came up against a Cape cobra, a snake considered to be one of the most dangerous of all African cobras. The reptile slithered out of a disused meerkat den just metres from my feet and, seeing me, reared up and flared its neck. I stumbled backwards in surprise, my knees weakened by cataplexy. This might seem like the worst situation in which to lose muscle tone, but in 2002, Dutch researchers came up with an idea that might explain my reaction to the cobra. Cataplexy could be an atavism, a biological trait thought to have been lost in the mists of evolutionary time that can suddenly pop up in the present.

In the nineteenth century, the study of embryonic development focused attention on this phenomenon. Species as different as a fish, a salamander, a tortoise, a chick, pig, cow, rabbit and human all start life with roughly the same body plan: an ET-like head, large eyes, a curving spine and curling, question-mark of a tail. It is only later on that the first signs of differences begin to appear. These similarities in early development reflect a hard-wired embryonic potential that may explain a multitude of peculiar natural phenomena, like the hind limbs on a humpback whale caught off Vancouver Island in 1919. 'The total length of the leg before it was cleaned of blubber and flesh was ... about 4 feet, 2 inches from the body,' noted one of the whalers. More recently, experimental biologists have been able to breed chickens with teeth, reactivating a genetic pathway that has been passed silently from one feathered ancestor to the next for some 70 million years. But perhaps the most well-known of atavisms is the human tail. This is present in early human embryos but self-destructs during normal development. If that doesn't happen, the upshot is a tail a few centimetres long extending from the base of the spine and winding up somewhere between the buttocks. Most parents who give birth to babies with tails understandably seek a surgical remedy.

Charles Darwin summed up the potential of an embryo rather nicely in *Variation of Animals and Plants Under Domestication* published in 1868. 'Besides the visible changes which it undergoes, we must believe that it is crowded with invisible characters ... separated by hundreds or even thousands of generations from the present time: and these characters, like those written on paper with invisible ink, lie ready to be evolved whenever the organisation is disturbed by certain known or unknown conditions.'

What if cataplexy is one such character? Dutch neurophysiologist Gert van Dijk hit upon this idea in a rather serendipitous manner. For many years he'd been interested in syncope, where a sudden drop in blood pressure causes a faint. 'Around 40 per cent of people will experience fainting at least once in their lives,' he says. This can be caused by a problem with the heart or simply by standing up too quickly. But much more interesting, from van Dijk's perspective, were those swoons triggered by an emotional response, as happens when someone topples over at the sight of blood or a needle.

In the hope of figuring out what might be happening during such neurally mediated syncopes, van Dijk went in search of other animals doing something similar and began to read up on 'tonic immobility', also referred to as playing dead, playing possum, fright paralysis, mesmerism, fascination and bewitchment.

Whatever its name, tonic immobility is 'a condition exhibited by numerous vertebrates, in which an animal is rendered immobile when faced with danger', and the most common explanation for this puzzling behaviour is one of predator avoidance. 'It's easy to see how a small mammal, in the jaws of a predator, might benefit from playing dead,' says van Dijk. 'The predator is more likely to relax its guard,' he says.

In his *Illustrations of Instinct* published in 1847, Cornish naturalist Jonathan Couch assembled many stories of animals playing dead. Like the occasion that a cat carried a weasel to its house, the poor prey 'dangling from her teeth as if dead'. With the front door shut, the puss laid her victim gently on the step, 'deceived by its apparent lifelessness', and began to mew to be let in. Whereupon the weasel jumped up, 'struck its teeth into its enemy's nose' and made a sharpish getaway. Couch considered this playing possum behaviour to be instinctive. 'The appearance of death was not the fictitious contrivance of cunning, but the consequence of terror.'

As van Dijk immersed himself in the 'tonic immobility' literature, however, he hit a problem. When a human faints, the brain is starved of oxygen, its activity drops and the patient completely loses consciousness, says van Dijk. 'Coming round is the result of blood flowing back to the brain, rebooting its activity.' When an animal plays dead, by contrast, it remains conscious, keeping a check on its surroundings. Its muscles will sometimes twitch, its heart rate slows and when it recovers it does so quickly. Tonic immobility, he realised, is a poor model for syncope. 'But it did sound a lot like cataplexy.'

In collaboration with sleep specialists at the same university van Dijk spelled out the similarities between tonic immobility and cataplexy. There's the emotional trigger, the muscle paralysis and the consciousness. Then there's the observation that a small dose of antidepressants tends to reduce the duration of tonic immobility in animals and is also one of the best ways to control cataplexy. The sudden loss of muscle tone during a cataplectic attack 'may represent a recurrence (atavism) of the ancestral trait of "tonic immobility",' they wrote. Some 15 years after van Dijk ventured the hypothesis, it is still a very real possibility.

I do like the idea that I might be able to empathise with a weasel, though if cataplexy really is equivalent to tonic immobility, I can see how natural selection might have shut it down: those human ancestors able to run from a Cape cobra would surely have left more descendants than those with a propensity for playing dead. Fortunately for me, I didn't suffer a complete collapse in front of the snake. Stumbling excitedly to safety, I survived to tell the tale.

It was after becoming a parent that I learned of yet another non-humorous emotion that could bring on a fit of the gibbers. My boys would be squabbling over some complete irrelevance, I would try to mediate, they wouldn't listen. With my anger rising, my muscles would weaken. The stronger the emotion the more likely it would be to cause a collapse. Losing my body posture seems like reasonable comeuppance for losing control of my children and my temper. On the up side, it always stops them bickering.

In the spring of 2015, chair of Narcolepsy UK Matt O'Neill experienced a rather more noble anger-fuelled attack. It was mid-afternoon and he was in his kitchen, when he heard a rapid-fire, high-pitched chirping, like marbles spilling onto a stone floor. He looked through the window to see a pair of magpies repeatedly diving into the hedge at the front of his house, a hedge he knew to be home to a pair of song thrushes. The parents were putting up a brave defence of their nest.

'I took an instant decision to join them,' says Matt. But somewhere between making that decision, jumping up and opening the front door, he began to weaken, and as he stumbled outside he went down. 'I didn't even get round to clapping and shouting, but lay on the gravel drive as the lead magpie dived and wriggled into the hedge. I was really angry.'

By the time the magpie had picked up a chick in its bill, Matt had wrestled back some muscle tone and faltered into a precarious squat. As the magpie attempted its getaway, Matt lurched towards it, landing cheek-first on the lawn. The baby song thrush fell from the magpie's startled jaws and landed on the ground beside him. With the thought that he'd thwarted the attacker, Matt was drowned in another wave of cataplexy. 'I thought how similar I was to the chick, stuck on the ground, helpless, unable to move, unable to help.' The mother song thrush, perched in a tree above him, was chirping away, clearly distraught. Eventually, Matt recovered and picked up the chick. It was warm, its heart beating rapidly, and he showed it to the mother before reaching into the hedge to return it to its siblings.

Matt has some even more unusual triggers, ones that none of the 100 or so patients in the cataplexy survey mentioned. He was at home when he first discovered smells could cause him to plummet. His wife, who had just varnished her nails, passed him in the hallway and he fell to the floor. 'It's a weapon of mass destruction,' he says. Indeed, the volatile chemicals in nail polish or nail polish remover cause him untold trouble on trains. If a woman so much as opens the lid and brings out a brush, Matt will slump in his seat and his jaw will fall open. 'I've learned not to fight it now,' he says. In the event that his head topples onto his neighbour's shoulder, he just goes with it.

Cosmetics are not the only smelly triggers. 'I'm still collecting them,' says Matt. There's trichlorophenylmethyliodosalicyl, or TCP, an excitant he discovered while out walking his dog in the village. 'A little old lady was walking down towards me and I smelt it,' he says. 'She took me out completely.' Matt has learnt to aim in the safest direction. He ended up hanging on some railings.

In recent years, there appears to have been an increase in the use of seaweed-based fertilisers on the beautifully manicured lawns of Chalfont St Peter. This is not something Matt would have ordinarily noticed, except that passing one such lawn on the way to the station, the oceanic odour hit him. 'I had to wedge myself up against a tree,' he remembers, a delay that caused him to miss his train. It's possible that the seaweed smell roused a potent memory. 'I think it comes from swimming through lots of kelp in the chilly Atlantic in Ireland as a kid. I hated the smell,' he says.

\* \* \*

Back in 1924, Olivia dutifully took her anti-epileptic medicines, but still found herself floored by humour. In May that year she returned to Queen Square, but came under the care of a different neurologist, a young Australian-born physician by the name of William Adie. This was a stroke of luck because Adie had recently become fascinated by narcolepsy, and clearly recognised the difference between 'attacks of irresistible sleep' and the 'curious attacks' triggered by emotion. Olivia became the first of 30 patients he assembled in a thesis he published in 1926, the first really accurate description of the narcoleptic condition, and the first use of the term 'cataplexy'.

'When I laugh I cannot stand,' Olivia told him. Adie wrote down her words verbatim. 'If anyone tells me a joke and I see it very much my eyes go misty, I feel funny in the head, and if I don't sit down I fall down.' The doctor got Olivia and her mother to describe one of these episodes in more detail. 'She opens her mouth, her head falls forward, the arms fall to her side and the knees give way,' he noted. 'She does not lose consciousness but is unable to move or speak; it lasts about half a minute; she never sleeps in these attacks.'

With my brain scans showing nothing untoward but my collapsing fits continuing, I too returned to Queen Square. I told my consultant that my collapsing fits continued and how I'd reacted to Rob Andrew's drop goal. I had also done a bit of digging in the library and I had scribbled the word 'narcolepsy' in a small, blue, ring-bound notebook I carried to the appointment, hoping that this then-alien word might mean something to a consultant neurologist. He batted away my suggestion, still set on the idea that I was experiencing 'reflex epileptic attacks'.

Then he announced he would like to see one of these fits, as if this would be easy to arrange. I imagined a laboratory, my scalp wired to a computer and an overdose of doctors and nurses, all on tenterhooks, waiting for me to laugh. It didn't sound funny in the least. It could never work. I politely shared my concerns.

If Adie had been in the room with us, he might have voiced similar doubts. 'Few detailed observations have been made during these seizures, because deliberate efforts to induce an attack do not often meet with success,' he'd written, 'and if an attack does occur, it is likely to be almost over before the examiner has recovered from his surprise.' This simple fact – that it's very, very hard to study cataplexy – helps explain why we are only now beginning to piece together what is going on in the brain during an attack.

But unaware of just how tricky it would be to catch me mid-fit, my consultant persisted, asking me whether there was any really reliable trigger. I thought for a moment and realised that his plan might just work after all.

'There is one thing,' I told him.

* * *

On Friday 13 October 1995, I walked into the National Hospital for Neurology and Neurosurgery with Zaid, the one person capable of causing a collapse to order. Just as Olivia had the good fortune to come under the care of William Adie, I was now being looked after by David Fish, a young consultant who had obviously heard of narcolepsy and cataplexy, because he'd already pencilled the diagnosis into my notes. Zaid's brief was as follows: to wait until I was wired up and the cameras were rolling, then make me fall over laughing. It was asking a lot, as I knew I was more likely to come under the muscular spell when in private and amongst friends. But Zaid likes a challenge.

A team of nurses began to fix electrodes onto my scalp, parting my hair, dabbing at my skin with an alcohol-infused pad, applying a gel and gluing wires onto my head. There were supposed to be more than 20 such wires connecting me to a computer, with others measuring my heart rate and muscle tone around my body, but Zaid began to lark around before the job was complete, tilting his head at me with mock concern, raising an eyebrow, lowing gently. I slumped into the high-backed chair, with only half of my wired dreadlocks in place.

Once I'd recovered and the nurses had completed their job, Zaid began to produce a series of 'props' from his bag, winking theatrically as he did so, extinguishing any lingering concerns over my ability to gibber in this formal, medical setting. His *pièce de résistance* was a photograph. He held it up with its back to me, tantalisingly out of reach, ramping up the anticipation. Then, all of a sudden, he turned it round to reveal four boys in evening dress, aged about 13, doing their utmost to look grown up. Three of them (one of whom was Zaid) just about pulled it off. The fourth (me) was still a couple of years off puberty and looked frankly ridiculous. It wasn't very funny, I'll admit, but the combination of anticipation, surprise and irreverent, only slightly funny humour was easily sufficient to trigger a full-on slump. With Zaid's assistance, I served up 13 attacks in the space of three hours, the longest of these lasting for 16 seconds.

As I sat there chatting to Zaid, my brainwaves zipped up and down as would normally happen during quiet wakeful-ness. When he made me laugh and my head dipped onto my chest, the readings coming from my muscles vanished, the senior technician reporting 'marked attenuation' a couple of seconds before each head drop. But my brainwaves motored on as if nothing had happened, confirming my own testimony that I was not asleep but had remained conscious throughout.

What is going on inside the brain during one of these attacks? Since my Zaid-assisted performance in 1995, there have been several efforts made to get an idea. Owing to the fact that strong emotions such as humour, elation, surprise and anger can trigger cataplexy, the attention has focused on the amygdala, two almond-shaped clusters of cells buried in the middle of the brain that play a key role in the processing of memory, decisions and emotions.

Around ten years ago, researchers in Switzerland designed an elegant experiment to take a look inside the brain at the point of one of these cataplectic attacks. They doctored photographs, inserting a comic version of each. In a photo of a dozen sheep grazing on a hillside, for instance, they inserted an extra sheep in the foreground and then a fox doing a leapfrog over it, its hind legs wide apart, its white-tipped tail flying through the air with what looks like a grin on its face. In another, they took a picture of several men being patted down by police and added a small dog to the line, its hind-legs straddled and fore-paws on the wall. The researchers then recruited a bunch of volunteers (some healthy, some with narcolepsy and clear-cut cataplexy), stuck them in an MRI scanner and showed them sequences of images, the original for three seconds, before suddenly flipping to the humorous version.

There were some telling differences. When a non-narcoleptic sees the humorous element, there is a surge in activity in the hypothalamus and a moderate response in the amygdala. In narcoleptics, by contrast, the hypothalamus remains ominously silent whilst the amygdala goes berserk. Clearly, there is something peculiar going on.

* * *

Massimo Zenti was ten when he first developed cataplexy, and the sensation was so scintillating, so lucid that it seemed to demand a fantastical explanation. 'None of my friends laughed and then melted like an ice lolly in the sun,' he says. 'I started to think I had superpowers.' Suspecting that 'maybe my brain can do something that the others can't', Massimo came up with an exciting idea. 'I started to think that I could move objects.'

'I remember concentrating on a bottle on a table about five metres away,' he says, closing his eyes and focusing all his attention. 'I can move it. I can move it,' he told himself. When he opened his eyes, the bottle was exactly as it had been, but these early telekinetic failures didn't put him off. 'I thought it was just a question of learning,' he says. 'I needed to train.'

So intense was Massimo's training, so hard did he concentrate and so excited was he at the prospect of success that his efforts would sometimes climax in cataplexy. Although he never managed to move an object, he did develop something surprisingly close to a superpower: he learned how to bring on a cataplectic attack to order. 'I begin by closing my eyes,' he says. Then he stops breathing and forces air into his Eustachian tube, the narrow passage that links the back of the throat to the middle ear, 'as divers do to equilibrate the pressure'. Finally, he rolls his eyeballs up and back in their sockets, almost as though he is looking backwards into his brain.

'Concentrate on your ears,' Massimo adds, before closing his eyes and showing me how it's done. Within seconds, I can see that the muscles around Massimo's mouth are twitching and I recognise the early signs of a cataplectic seizure. He doesn't take the fit any further, resurfacing, but I can see that it takes a second or two for his eyes to roll back into focus. Massimo's mouth turns into a broad smile. I have a go myself and instantly recognise the first whimperings of cataplexy. I am in no doubt that were I to practice, I too could quickly master Massimo's method.

'Once I realised that I could not move things but I could bring on the cataplexy, I began to play with it,' he says. 'I worked out how to bring myself out of an attack.' This was something he perfected while watching *Dragon Ball*, a Japanese TV series from the 1980s based on a manga cartoon strip of the same name. Goku, the protagonist, is a monkey-tailed boy with superhuman strength who goes in search of seven wish-granting 'dragon balls'. Massimo and a friend Andrea were watching an episode in which Goku turns from his normal state into the turbocharged Super Saiyan 3 for the first time. 'For a boy of my age, it was a super-exciting transformation,' he says, exactly the sort of thing that would trigger a collapse.

On this occasion, however, Massimo applied his understanding of cataplexy in reverse, just to see if he could fend off an attack. 'Keep your eyes open and still. Concentrate on continuing to breathe, steadily and slowly. Try to keep your ears open,' he advises.

'This is to go even further beyond,' shouts Goku just moments before turning from Super Saiyan 2 into Super Saiyan 3. 'Aaaagh!' he screams for minutes on end, his pitch gradually rising like a jet engine in terminal decline. Andrea turned to him, unable to understand why the transformation was not triggering a fit. 'Look Massi, look! Goku's transforming,' he said. 'Don't you care?' Massimo did care of course, it's just he'd found a way to suppress cataplexy.

Over a decade later, when Massimo told his sleep specialist that he could do this, entering and coming out of an attack at will, he wasn't believed at first. But Giuseppe Plazzi, the neurologist in Bologna who has argued that Dante suffered from narcolepsy, soon realised that if Massimo was right, he could be a useful research subject.

Which is how Massimo came to be sitting in a comfortable

chair under the care of Plazzi's colleagues Christian Frances-chini and Vincenzo Donadio, a series of fine tungsten needles inserted into a nerve in his leg, electrodes on his chest to pick up his heartbeat and a belt around his ribs to measure his breathing. With everything set, the doctors asked him to perform. Massimo closed his eyes and went through his tried and tested routine. 'I did it,' he says. 'I went completely paralysed and could bring myself out.' Everyone was duly impressed.

This study revealed that Massimo had entered a perfectly normal cataplectic paralysis. But while all the signals to his muscles cut out, the nerve in his leg suddenly went into over-drive. What was it doing? Most likely causing a constriction of blood vessels and a rise in blood pressure, says Donadio. These are precisely the sorts of changes we see in animals that are playing dead.

It's a useful skill to have. If Massimo wants to go to sleep, all he needs to do is to bring on a cataplectic attack and keep himself there. It doesn't take long before sleep takes over. He can use it to kick-start a lucid dream. With the cataplexy underway, he just begins to think of the subject matter, like being a world-class footballer, for instance, and when the sleep comes he's there.

\* \* \*

With a bunch of cataplectic attacks captured on film, the diag-nosis of narcolepsy and cataplexy was looking increasingly likely for me. But in order to be sure, sleep specialists will normally insist on two sleep tests to get a better idea of what is going on inside the brain during sleep, so that's what I did next.

# 4

# Stages of sleep

*'The sleeping brain was engaging in activities
that could be observed by instruments, and of
which the sleeping mind knew nothing.'*
Kenton Kroker, *The Sleep of Others*

Hans Berger could do nothing as the huge field gun rolled towards him.

In 1892, the 19-year-old German had enlisted for military service in Würzburg. One spring morning, while pulling heavy artillery for a training session, Berger's horse suddenly threw him to the ground. He watched, helpless and terrified, as the rolling artillery came towards him, only to stop at the very last minute.

At precisely the same moment, Berger's sister – several hundred miles away in his home town of Coburg – was struck by a premonition, an overwhelming sense that something tragic had befallen her brother. She begged her father to send him a telegram to make sure he was okay. Berger was stunned by the coincidence. 'It was a case of spontaneous telepathy in which at a time of mortal danger, and as I contemplated certain death, I transmitted my thoughts, while my sister, who was particularly close to me, acted as the receiver,' he wrote.

'Berger never forgot this experience, and it marked the starting point of a life-long career in psychophysics,' says historian

David Millet of this episode. Berger began to study the brain and the electrical signals it gave off, determined to make sense of what he called 'psychic energy'. In a sense, he succeeded. His efforts to record the small electrical signals that escape from the brain and ripple across the scalp have given us one of the key tools for studying sleep, the electroencephalogram (EEG), or, as Berger described it, 'a kind of brain mirror'. It was this device that I'd been wearing when Zaid triggered my cataplexy in 1995.

In 1929, Berger published his discovery. As others looked to replicate Berger's work, it soon became apparent that the EEG revealed electrical activity during sleep too. Based on the EEG signature, researchers could show that there were several different stages to sleep, and the sequence and timing of them underpins the diagnosis of many sleep disorders. But in the first few decades of using the EEG, there was one stage of sleep that nobody noticed, one that turns out to be crucial to the diagnosis of narcolepsy.

\* \* \*

During a long train journey in the 1940s, Robert Lawson (a physicist at the University of Sheffield in the UK) made an interesting observation. He was sitting in a carriage with a young man and his wife, and as the train rattled along both his fellow travellers fell asleep several times. Lawson began to collect data, recording the frequency of blinking when his fellow passengers had their eyes open and closed. 'The subjects were quite unaware that they were under observation,' he wrote in a short letter to *Nature* in 1950. With their eyes open, both the man and the woman blinked roughly once every two seconds. When they closed their eyes, Lawson could see their eyelids twitching at the same frequency for a time. Then, quite suddenly, the blinking stopped altogether, suggesting to Lawson that the transition from wake to sleep was not gradual but sudden.

When Nathaniel Kleitman – then 'the most distinguished sleep researcher in the world' – read this casual observation he set a young graduate student the task of probing the Lawson observation. Eugene Aserinsky had majors in social science and Spanish, had trained to be a dentist and once worked in the US army as an explosives handler. On Kleitman's instructions he buried himself in the literature on blinking, with the aim of becoming 'the premiere savant in that narrow field'.

Long after his retirement, Aserinsky recalled the prospect of studying the movements of eyelids during sleep as a 'research turnip' and 'about as exciting as warm milk'. But 'painstaking, diligent exploration of minutiae will frequently lead to the "golden manure" phenomenon whereby there is a rewarding result,' he wrote. Forgiving the strangeness of Aserinsky's metaphors, this is exactly what happened in the case of his work on blinking.

Based on Lawson's observations on the train and preliminary investigation of sleeping babies, Aserinsky guessed that the eyes might be doing something interesting during sleep. 'I had good reason to hope, if not expect, that an examination of eye movements would yield some unrevealed aspect of brain function,' he wrote.

Measuring eye movements during sleep might seem like a trivial exercise today, but back then it wasn't. Even with a basic electronic device, the recordings were hardly reliable. The instrument 'spontaneously spewed forth pen movements even when no subject was attached'. This made it difficult to know what was a genuine eye movement and what was the machine playing up.

As Aserinsky tinkered away with his equipment, he was often joined in the lab by his young son. Armond Aserinsky is now in his 70s, a retired clinical psychologist living in Palm Harbor, Florida. 'The building was old and dark,' he recalls. 'It was like something out of the horror movies of the 1930s.' This might have put off an ordinary eight-year-old, but for young Armond these were exciting times. He lived on campus with his parents and his sister, with the University of Chicago as his playground.

On one occasion, Armond remembers heading down to the chemistry building with a bunch of boys – the children of other academics – hoping to secure some test tubes for their chemistry sets. At the entrance to one lab, they were met by a handsome gentleman in his late 50s, who dug out some used equipment and sent the children off with a selection of glassware. It was only years later that Armond realised the significance of this encounter. The scientist had been none other than Harold Urey, who had won the Nobel Prize for Chemistry in 1934 for his discovery of the hydrogen isotope deuterium, and who, when Armond came knocking, had been in the midst of the famous Miller-Urey experiment to explore the origins of life.

In spite of its horror movie-like qualities, Armond's father's laboratory also had its attractions. There were cupboards and drawers filled with obsolete equipment that he was free to play with: retort stands, clamps, tripods, scales, weights, even the occasional brass microscope. 'I might take some roller skates along,' he tells me. Then, if his father was busy with something particularly tedious, Armond could always head outside for a skate.

More often than not, however, Eugene Aserinsky would involve young Armond in his research, bouncing ideas off him, asking him to read through a manuscript and using him to calibrate the equipment. 'Electricity is coming out of your brain,' he explained to his son, 'and this machine is going to measure it. It will be interesting to see what's produced when you're asleep.'

Armond remembers one session in particular. It was the afternoon and the eight-year-old was standing in the sleep room, a chamber furnished only with a cot-like bed, with an intercom as its sole means of communication with the outside world. He had electrodes on his scalp and his eyelids, with wires that would transmit his brainwaves and the movement of his eyeballs to the recording equipment outside. Instructed to lie down and try to sleep, Armond – like a good boy – did as he was told.

An hour or so into his nap, the readout suggested that his eyeballs had suddenly gone crazy, jerking rapidly from left to right. 'My father woke me and asked me what was going on.' Armond had been dreaming. 'He seemed very interested in this and asked me what the dream had been about.' More than 65 years later, Armond can still remember his answer. 'There was a chicken walking through a barnyard.'

As Aserinsky's studies progressed, his sleeping subjects appeared to enter a categorically different state. Poring over the reams of paper that had spooled out of the polygraph ('up to a half-mile length of paper per sleep session'), there were times when the brain signals looked almost indistinguishable from the signals during waking, yet their eyeballs were jerking and the subjects were obviously still asleep.

'There was no doubt whatsoever that the subject was asleep despite the EEG suggesting a waking state,' he wrote. This was what became known as rapid eye movement, or REM (usually pronounced 'rem' rather than R.E.M).

When one of the volunteers experienced a 'REM hurricane', with such violent eye jerks that the pens leapt off the sheet of paper, Aserinsky witnessed the frantic eye movements himself and heard the volunteer mumbling in his sleep. If he woke his subjects up during these periods and asked them what they remembered they often mentioned 'remarkably vivid visual imagery'.

The link between eye movements and dreaming had been mooted before. 'What was entirely unanticipated by myself and all my predecessors in sleep research was that there exists a unique stage of sleep which reappears periodically in a near cyclic fashion, and has attributes different from either waking or traditional sleep.' Aserinsky and Kleitman wrote up these findings for *Science* in 1953, though they were so focused on the eyes that they failed to notice the very significant fact that REM is also accompanied by a complete loss of muscle tone throughout the body, most likely to prevent you acting out your dreams. This observation, made by French neurophysiologist Michel Jouvet and his colleagues several years later, is pertinent to several sleep disorders, (including the cataplexy that is so often a feature of narcolepsy), and one that we will pick up on in due course.

* * *

The discovery of REM inspired a flurry of research. Aserinsky did not stick with Kleitman but headed off to the University of Washington in Seattle to study the effects of electrical currents on young salmon. But psychology graduate William 'Bill' Dement, who had recently worked his way into Kleitman's circle and had helped Aserinsky with his work, was excited by the possibility that REM might be an objective way to study dreaming.

In 2017, I travelled to California to meet Dement. Now in his late 80s, he retired several years ago, but still lives in a large, old and strangely tall house nestled in a leafy neighbourhood on the edge of the Stanford campus. If the Dement household had been standing on a hill, above a motel, I might have thought I'd arrived on the film set for *Psycho*.

Dement's office, a large, shed-like structure attached to the main house, is not unlike a scout hut. The walls are wood-clad and covered with framed posters, photographs and miscellaneous memorabilia from an illustrious career in sleep medicine. Dement's desk is a picture of organised chaos. There is a large monitor, a laptop, a printer and a telephone, all of them contributing to a cascade of wires. There's a pigeon-hole-like filing system, a pile of scientific papers, a stack of novels, a couple of boxes containing material pertaining to the history of the National Center on Sleep Disorders Research and, incongruously, a water gun.

An electric strip heater blasts out a fuggy heat, precisely the kind that sends a narcoleptic to sleep. It's not exactly well-lit either, a greyish cloud-filtered light struggling through a pane of stippled glass. In fact, Dement, with his thick Aran-style jumper, ivory moustache and pearly head of hair is almost the brightest object in the room. He gets to his feet.

'Do I know you? Have we met before?' he asks. I've seen so many photos of Dement, watched videos of him giving other interviews, I feel like maybe we have. We take our seats, Dement in a rather beaten and incredibly squeaky rocking chair and me next to the water gun. 'It's for when students fall asleep in class,' he explains.

Within a few years of the discovery of REM sleep, Dement and Kleitman had come up with an EEG-based description of a normal, healthy night's sleep. The precise definition of the different stages has changed over the years, but they are currently characterised as follows:

From waking, the brain transitions to a light phase of sleep referred to as Stage 1, in which the eyeballs start drifting in their sockets and the EEG signals slow. In Stage 2, the eyeballs come to a standstill and the EEG trace is characterised by rapid bursts of activity known as 'sleep spindles' and 'K complexes'. In Stage 3, the brainwaves slow further still. If the waves of Stage 2 sleep are comparable to the waves of a beach – 'small, uneven, of varying shapes' – then the waves of Stage 3 sleep are like 'the great ocean swells,' wrote Dement in *The Promise of Sleep*. The sleep spindles and K complexes are still there in Stage 3 sleep, 'but like the wind-blown waves running over ocean swells, they are harder to see'. All of these stages – 1, 2 and 3 – are collectively referred to as 'non-REM' sleep.

Then, all of a sudden, the brain resurfaces and passes, as if through some cognitive portal, into the REM state. The 'hurricane' of ocular activity might last a matter of minutes, before the brain returns to the relative calm of the non-REM realm and the entire cycle starts again. The duration of this cycle – from the start of Stage 1 to the end of REM – typically lasts around 90 minutes and repeats throughout the night, though the duration of Stage 3 sleep tends to decline and the duration of REM tend to increase with every passing cycle.

'I believe the study of sleep became a true scientific field in 1953, when I was finally able to make all-night, continuous recordings of brain and eye activity during sleep,' wrote Dement. 'For the first time, it was possible to carry out continuous observations of sleep without disturbing the sleeper.' From that point onwards, the best way to reveal pathological patterns of sleep was to bring a patient into the laboratory, wire them up to an EEG and get them to perform an over-night sleep study.

Just days after Zaid had triggered a cataplexy attack, I returned to the National Hospital of Neurology and Neuro-surgery for a night away from home. I was refitted with the electrodes and shown to a room. There was a television, a video player and a selection of movies, a luxury I didn't have access to at university. I put on a film and climbed into bed. Before too long, the sleep technician told me to settle down to sleep, switched off the lights and gently closed the door.

During the overnight sleep study, the clinicians are looking for unusual nocturnal behaviour and EEG readings that deviate from the standard sleep cycle. If a patient has sleep apnea, for instance, the chances are they will stop breathing over and over throughout the night (see chapter 6). If sleepwalking is an issue, it could well show (chapter 8). Sleep paralysis is easy to detect (chapter 9). Insomnia will be self-evident too (chapter 10). If there are periodic limb movements, the sleep techs are going to spot the patient twitching or thrashing about in their sleep (chapter 11).

I awoke from my overnight sleep study just before 8 a.m. I had not shown any obvious signs of any of these other sleep disorders, but I had experienced the crazy dreams that are typical of narcolepsy.

\* \* \*

In the years following the discovery of REM sleep in 1953, Dement (who bought into Freudian thinking at this time) began to frame REM as some kind of safety valve for the mind, allowing the brain to blow off suppressed thoughts or memories. The hypothesis made two allied predictions: those with mental illness would have little or no REM sleep; and depriving healthy volunteers of REM would result in mental instability.

As a junior intern at Manteno State Hospital near Chicago, Dement studied schizophrenic patients as they slept. Contrary to the prediction, they all experienced REM and they all reported dreaming. Undeterred, Dement began depriving healthy volunteers of REM, waking them up each time the EEG waves slipped into a wake-like state and the eyes began to flit in their sockets. Unsurprisingly perhaps, his subjects became anxious and upset, one quitting 'in an apparent panic', two so stressed they had to stop after four rather than five consecutive days, but there was no evidence that REM deprivation resulted in mental illness.

Dement did have sufficient data though to suggest that the human brain needs 'dream time' of around 80 minutes a night and if it doesn't get it – as in this early study – it attempts to catch up. After his subjects suffered their REM-deprived ordeal, they experienced an increase in the amount of REM during subsequent sleep. It certainly looks like REM is serving some vital physiological purpose crucial to the proper functioning of the brain.

People have been looking for meaning in their dreams for a very long time. In ancient Egypt, priests doubled as interpreters of dreams, and considered significant dreams to be divine messages. Greek mythology held something similar, with a bunch of gods and demigods (collectively known as the Oneiroi) responsible for fashioning dreams and nightmares in the minds of sleeping mortals. In the second century AD, professional diviner Artemidorus Daldianus travelled around Greece, Italy and Asia asking people to recall their dreams. Being subjected to the plodding 'and-then-ness' of someone else's dream can be a trying experience, but old Artemidorus doesn't appear to have found it so, bringing some 30,000 dreams together in his whopping five-volume *Oneirocritica* and offering suggestions of their meaning.

Roughly 1800 years later, Austrian neurologist and founder of psychoanalysis Sigmund Freud was doing something remarkably similar with his *Interpretation of Dreams*, working on the assumption that dreams must be a form of 'wish fulfilment' that reveals repressed and often sexual desires. Although few people now buy into such strict Freudian thinking, a recent study suggests that most still believe that there is meaning to be gleaned from dreams.

In fact, this popular pastime is probably a colossal waste of time and energy. In the 1950s, while Dement continued to pore over the superficial EEG signals of his subjects and patients in the hope of finding out the function of REM and hence dreams, Michel Jouvet at the University of Lyon made an intriguing discovery suggesting that dreams might not be the raison d'être of REM. When he stripped back the brains of cats, removing the thick layer of thinking cortex where most of the dreaming action is thought to take place, the animals still slept perfectly well, with a regular cycle of both non-REM and REM. REM (or paradoxical sleep, as he insisted on calling it, because the brain appeared to wake while the body entered a state of muscular paralysis) had its origins in an ancient region of the brainstem called the pons.

This observation is at the heart of an article published in the *American Journal of Psychiatry* in 1977 by Allan Hobson and Robert McCarley, both psychiatrists at Harvard Medical School. Their rather clunkily named 'activation-synthesis hypothesis' proposed that REM begins with some kind of 'activation' in the pons, a content-free pulse that only acquires meaning through the 'synthesis' of vivid imagery, crazy plot-lines and intense emotions as it ricochets through the cortex. This sequence of events effectively demotes dreaming to something of a secondary, perhaps inconsequential after-thought. 'The process,' they wrote, 'casts serious doubt upon the exclusively psychological significance attached to both the occurrence and quality of dreams.'

In *The Promise of Sleep*, Bill Dement captured the essence of Hobson and McCarley's hypothesis with a sparkling analogy:

> It might help to think of a stained-glass window ... White light, which is a jumble of colors, enters on one side, but what comes out on the other side has a definite pattern of colors that is often very meaningful. Like the stained-glass window (which is a filter for light), the brain acts as a filter that imposes order on the random signals passing through it.

If this is how dreams come about, it is almost inevitable that they should contain some meaning. The white light (the REM signal from the pons) has, after all, passed through a stained-glass window (the cortex), a filter that differs from person to person, and so may illuminate very individual mem-ories, experiences and emotions. But the output is unlikely to resemble the kind of coherent artistry found above the altar in a church. In fact, extending Dement's stained glass analogy, it's probably more accurate to think of REM as a flickering strobe light and the cortex as layer upon layer of kalaedoscopic glass of variable colour and thickness. The light that emerges on the other side will certainly make patterns, but it is prob-ably a waste of time trying to read too much into what they might mean.

As it became increasingly clear to most clinicians that the interpretation of dreams was little more than storytelling, so research into REM sleep itself rather fell out of favour, at least for a time. The investigation of the non-REM stages of sleep, by contrast, proved to be a more profitable line of enquiry, one that has got us close to figuring out what sleep is actually for.

* * *

I have asked many researchers and clinicians about the function of sleep, and I almost always get a different answer. I make two related conclusions. First, there is no consensus. Second, sleep most likely performs more than just one function.

There have been many theories about the function of sleep over the last century. Some of them don't make much sense, like the suggestion that sleep might simply be a way to cope with total darkness. The sleep-to-cope-with-darkness hypothesis could conceivably come into play in humans but the existence of nocturnal species that function perfectly well at night and sleep during the day strongly suggests that the ultimate function or functions of sleep, the reason or reasons why it evolved, requires a far more fundamental explanation. Sleeping to hide from danger is not much more convincing. For if this were the function of sleep, top predators like lions wouldn't bother sleeping, and they do. A lot.

An allied idea that sleep evolved in order to save energy has more mileage – a hypothesis that Jerry Siegel, a sleep researcher at the University of California, Los Angeles, refers to as 'adaptive inactivity'. Sleep 'reflects the evolutionary advantage of being inactive when conditions are not perfect,' he says. 'If you have to continually spend energy even when it's not easily available you are going to leave fewer offspring or none at all,' he says.

Some people have countered the notion of sleep as adaptive inactivity by pointing out that a lot is going on in the brain during sleep and the amount of energy saved is minimal, in humans the equivalent of a piece of bread or a hotdog roll. 'My take,' says Siegel, 'is that saving a little energy is not trivial. If you could give half of the people on earth a piece of bread once a day they would do a lot better than the half that doesn't have a piece of bread.'

There's a lot to be said for this idea. It would help to explain why even closely related animals sleep so differently. The amount of sleep a species gets is simply a function of the amount of downtime it can afford. If the balance between activity and inactivity is wrong or sleep strikes at an inappropriate moment, the sharp scythe of natural selection comes sweeping through the crop.

Another intriguing explanation for the evolution of sleep is suggested by a chance observation in the box jellyfish *Chironex fleckeri*, a species infamous for its lethal sting. Less well-known is the fact that it has 24 eyes and appears to sleep. Box jellyfish are active predators and during the hours of daylight they are extremely mobile, typically covering around 200 metres an hour. At night, they basically stop. 'During these periods of "inactivity", the jellyfish lie motionless on the sea floor, with no bell pulsation occurring and with tentacles completely relaxed and in contact with the sea floor,' wrote Jamie Seymour, a biologist at James Cook University in Cairns. A small disturbance – like a light or a vibration – 'causes the animals to rise from the sea floor, swim around for a short period, and then fall back into an inactive state on the sand,' they reported. To Seymour and friends, this looked a lot like sleep, and they put forward an adaptive inactivity-like explanation. In the dark, when vision is limited, 'it makes a lot of sense to become inactive, decrease your energy used in locomotion and divert it to growth'.

When evolutionary biologist Lee Kavanau read about jelly-fish sleep, however, it leant weight to an idea he'd been mulling over for several years. Eyes – for all their obvious evolutionary advantages – require huge neurological processing. So, in a paper published in the late 1990s, he argued that the evolution of eyes paved the way for the emergence of two distinct states of vigilance. One – wakefulness – allowed the animal to focus on the analysis of complex visual information and the split-second making of decisions. The other – sleep – became the brain's opportunity to process information without being over-loaded by the senses. Species with little or no vision seemed to need less or no sleep at all, whereas the presence of eyes (even in an invertebrate like the box jellyfish) would place categori-cally different demands on the brain, he argued.

There are other not-too-dissimilar proposals for the func-tion of sleep. It could, for instance, be for 'brainwashing', a way to purge a load of pointless information accumulated during the hours of wakefulness. In 2003, biologists at the University of Wisconsin-Madison developed this notion. The brain is so busy making connections when an animal is awake, they argued, that sleep is needed to pare back on some of this neurological noise.

More than a decade and a bundle of research papers later, there is compelling evidence that some kind of neuronal editing does indeed take place during the non-REM stages of sleep. In a paper published in 2017, researchers collected slivers of brains from waking and sleeping mice and visualised the neurons, scoring the size and shape of almost 7,000 synapses. Sleeping mice had a significantly reduced area for neuronal cross-talk, but crucially this somnolent downsizing did not occur across the board. For some synapses, particularly the larger, more crowded ones, sleep did not result in shrinkage, leading the researchers to speculate that these might encode important memories that need to be preserved.

Researchers elsewhere have been busy pinning down the molecular processes that might result in such synaptic scaling. In mice engineered without a particular gene (*Homer1A*, if you must know), synaptic pruning does not occur as it should. In memory tests conducted on mice incapable of pruning, they prove to be less decisive than normal mice.

In addition to this subtractive process, there's also evidence that sleep can have an additive effect, with certain synapses, circuits and memories being strengthened. In 2014, for instance, researchers demonstrated that the neurons of mice that had been through a process of learning branched during non-REM sleep.

There is evidence too that non-REM sleep may be a time for the brain cells to carry out important housekeeping duties, replenishing stores of neurotransmitters and calcium, for instance. Brain cells also appear to shrink somewhat during non-REM sleep, allowing more room for cerebrospinal fluid to percolate and wash away toxic metabolic waste.

In spite of Dement's best efforts, we haven't made the same headway with REM. 'I'm adjusted to that,' he says. 'I created a whole new clinical discipline. I'll take that.' Indeed, we have Dement to thank for both the overnight sleep study and what's called the multiple sleep latency test, or MSLT, to get a handle on daytime sleepiness. These two tests underpin the diagnosis of most sleep disorders, including narcolepsy.

\* \* \*

In 1960, Chicago-based psychotherapist Gerald Vogel published an extraordinary paper. Steeped in the Freudian interpretation of dreams, he was interested in exploring the idea (then prevalent, now batty) that narcolepsy might be some kind of escape mechanism, a means to avoid anxiety perhaps, a nifty way to repress inappropriate feelings or, in the case of one woman, as 'a defense against the guilt of actual incest with her father'.

One of Vogel's patients, 'an intelligent 42-year-old married Negro male shipping clerk', recounted how he had started to experience 'irresistible periods of sleep' at the age of 13, 'in quiet, monotonous classes such as history, literature, and study periods'. In other lessons where there was physical activity, as in PE, or constant mental demands, like in maths, he remained alert. But it soon got so bad that 'I would go to sleep while walking to school,' he said. 'When I reached the curb, stepping off would revive me.'

After 20 years living like this, dabbling with a series of menial jobs and ending up in an occupation 'beneath his intellectual status', the clerk was diagnosed with narcolepsy and put on a course of amphetamines. It was transformative. 'From a fellow who came home, read the paper and went to sleep, who had no interest to do anything, I became alert, active, interested,' he told Vogel. 'I started drawing.'

Reading between the lines of this case study, I imagine Vogel raising a Sigmundesque eyebrow at this point, his interest piqued by what these drawings might reveal. He was in for a treat. 'His paintings are almost exclusively graphic representations of his feelings and fantasies about women,' wrote Vogel, opening up an almost infinite canvas for Freudian interpretation. 'The sleep of narcolepsy may provide some kind of gratification of these fantasies which are unacceptable to waking life.'

Though this is easily the strangest idea about narcolepsy that I've ever come across, it's worth sticking with Vogel for a little longer. His working hypothesis – 'that the narcoleptic patient makes use of his sleep for the projection of a fantasy which is gratified in a dream in a way unacceptable during waking life' – led him to predict that 'the narcoleptic may dream very quickly after the onset of the sleep attack'.

Vogel got the poor shipping clerk to stop taking his amphetamines and had him wired up to an EEG. When asked to sleep, the patient did so within a minute. 'Simultaneously with the onset of sleep he was almost certainly dreaming', and in just three minutes the EEG trace showed he was caught up in one of Aserinsky's REM hurricanes. When Vogel woke the patient up, he was thrilled to discover that he had indeed been dreaming, a Borges-esque fantasy involving a doctor in a suit, a hall leading to a succession of frill-curtained rooms, one of them 'a dormitory with ladies'. When the patient repeated the procedure a second time, he was dreaming again almost immediately. 'Narcoleptics sleep to dream,' was Vogel's succinct proposition.

With hindsight, much of Vogel's reasoning seems preposterous. But his study does contain one observation that has stood the test of time: people with narcolepsy enter the REM state far more quickly and far more readily than most.

In a follow-up investigation, Vogel's colleague Allan Rechtschaffen teamed up with Bill Dement and others to publish a more robust description of the EEG patterns of nine people with narcolepsy (one of them the 'Negro shipping clerk'). They found 'sleep onset REM periods' (or SOREMPs, as they are known in medical circles) to be very common, so distinctive in fact that their occurrence 'may be a diagnostic aid'.

\* \* \*

In the early 1960s Dement had moved to Stanford, and in 1970 he and his wife Pat had been appointed as resident fellows of the university's first ever mixed-race dormitory. In the dorm, everyone got along beautifully, but with racial tensions running high across the wider campus, Dement encouraged the students under his care to stay indoors, and he began giving lectures on sleep to pass the time.

These proved immensely popular, so much so that students from other dormitories began to attend. With the in-house students feeling a little put out by the intrusion, Dement offered to give a course on sleep for the whole university in the next quarter. 'To my amazement ... a huge number of students signed up,' he says. 'The number 600 comes into my mind.' Without a room to accommodate such a large audience, Dement delivered the first lecture in the course from the pulpit of the Stanford Memorial Church.

'It was very uncomfortable,' he says, and he half expected a lightning bolt to strike him down for such heretical behaviour. But the course proved so popular that it quickly became something of a fixture in the Stanford calendar, with students from every field imaginable coming along to hear Dement speak on sleep and dreams.

At around the same time, Dement set up a rudimentary sleep lab in the basement of the dormitory and began to collect data from some of the students. It was in this setting that he and his colleague, Mary Carskadon, came up with the MSLT.

It is daytime; the curtains are drawn; the patient is plugged in to an EEG, tucked up and told to sleep. After 20 minutes, the curtains are opened and the patient is told to get up. The whole procedure is repeated five times, allowing the sleep doctors to calculate the average time it takes the patient to fall asleep. Dement and Carskadon coined the phrase 'sleep latency' to define the interval between the start of the test and the first flickerings of sleep on the EEG monitor. A sleep latency of less than five minutes qualifies as 'severe' hypersomnia, five to ten minutes as 'troublesome', ten to 15 as 'manageable' and 15 to 20 as 'excellent', meaning that excessive daytime sleepiness is not an issue at all.

The morning after my overnight sleep test at Queen Square in 1995, I was subjected to the MSLT. I ate breakfast, watched some TV and then the sleep technician instructed me to go to sleep once more, a task I achieved with ease. Across the five trials, I was asleep with an average sleep latency of just 90 seconds, so according to Dement and Carskadon's MSLT scale I am a severe hypersomniac.

When two people with narcolepsy get together they will often size each other up by exchanging MSLT scores. I am always impressed that they have made a mental note of this figure at all, though I think I understand why. The MSLT is one of the few tests at which people with narcolepsy really excel. It is also an important validation of their condition, one that may have taken years or even decades to obtain.

As impressive as a 90-second sleep latency might sound, however, it's pretty average for someone with narcolepsy. When I posted my MSLT score on an online forum for people with narcolepsy, Lucy Tonge pipped me with 68 seconds. But the clear winner, with an average sleep latency of just 10 seconds, was Chloe Glasson.

The MSLT is not just a way to measure daytime sleepiness; it is also a great opportunity to look for those SOREMPs that are characteristic of narcolepsy. When someone with healthy sleep is subjected to a MSLT and is given five opportunities to snatch a 20-minute nap during the course of a day, they will most likely struggle to fall asleep. If they do, they are extremely unlikely to experience REM sleep in such a short window. SOREMPs in two out of the five trials is considered diagnostic of narcolepsy.

I experienced SOREMPs on four out of five trials, with plenty of the bizarre, vivid and convoluted dreams that are standard fare for narcoleptics. I was discharged with instructions to return a few days later and make an appearance at something called the 'Gowers' Grand Round'.

In 1949, William Gowers was described by one brown-nosed colleague as 'probably the greatest clinical neurologist of all time'. At the National Hospital for Neurology and Neurosurgery, his legacy lives on in the form of 'Gowers' Grand Round', the highlight of the teaching week at which medical students cram into a tiered lecture theatre to mull over a perplexing real-life case. On this particular day, I was to be that case.

At the appointed hour, I took to the stage in the Wolfson Lecture Theatre. A pomposity of professors lined the front row and the interrogation began, the consultants teasing out my medical history and symptoms for the edification of the students squeezed behind them. A still image of me sitting in a straight-backed chair appeared on the projector screen and someone pressed play. I was relieved to note that there was no soundtrack; I had no problem with the students gawping at this spectacle, but I would have been mortified if they'd been party to Zaid's wholly inappropriate banter. By the end, I was fielding questions from every corner of the auditorium. In my medical notes, there is a three-page write-up of my appearance at Gowers' Grand Round. There was no doubt in anyone's mind about the diagnosis. Given my exhibition of cataplexy, the abundance of REM during my overnight sleep test and the number of SOREMPs I experienced, this was a clear-cut case of narcolepsy with cataplexy.

\* \* \*

From Bill Dement's house, I make the short drive across Palo Alto to meet Emmanuel Mignot, Dement's successor as director of the Stanford Center for Sleep Sciences and Medicine. I am hoping he might be able to bring me up to speed with the current thinking about REM. 'Ah,' he says, acknowledging in a breath that nobody yet knows. 'That's a tough one. I can give you my opinion, but it's just an opinion.' He proceeds to talk me through an idea he published almost ten years ago. 'It's the only speculative paper I've ever written,' he says. 'But I still think it's pretty close to the truth.'

In Mignot's view, REM has all the hallmarks of being an ancient phenomenon, one that evolved way back in the mists of vertebrate evolution before the forebrain had a chance to expand. Perhaps, he suggests, REM was the primitive brain's way of getting some rest.

The idea has considerable appeal, not least because the REM state has its neurological origins in the pons, a primitive structure that is conserved across the vertebrate board. It also provides an explanation for one of the most baffling things about REM: why it should send most of the body's core physiological functions offline. The skeletal muscles shut down, the body temperature free-runs, breathing becomes irregular, the heart rate races, blood pressure rises and the blood vessels dilate. (These last three phenomena, incidentally, explain why REM is often accompanied by the erection of both the penis and the clitoris.) Mignot argues that REM sleep could have evolved to allow basic locomotor, sensory and thermoregulatory circuits to recover.

'During REM, you become a little like a reptile,' says Mignot. It was only later on in the course of evolution, when the cortex came to dominate the brain, that REM took on a secondary role, triggering the imagery and emotions we refer to as dreaming, he suggests.

Like Hobson and McCarley's activation-synthesis hypothesis, the way that Mignot sees REM turns dreaming into something of an evolutionary afterthought. But, he stresses, this doesn't mean that dreaming serves no purpose. In fact, because REM is such an active state, burning up as much if not more energy than quiet wakefulness, it's likely that it could have acquired some secondary function in the cortex-heavy brains of evolved creatures like humans.

There is no shortage of ideas as to what this might be. The extraordinary amount of time that the foetal and neonatal brain spends in REM suggests that it could have some role in neurological development. Since REM activity acquires meaning as it blazes through the cortex, it could be helping to lay down new memories. Alternatively, REM-state dreaming might be strengthening or pruning existing pathways as occurs in non-REM, or both.

I am intrigued by the observation that most dreams involve a degree of peril. Negative emotions are very common. Misfortune is a persistent theme. Aggressive interactions are a regular feature. Many dream scenarios are also in keeping with the kind of evolutionary environment in which humans evolved, with carnivores, snakes and spiders popping up with alarming regularity. By contrast, relatively recent, less animate objects like books, typewriters and computers make surprisingly few appearances in dreams. Those suffering post-traumatic stress typically replay their waking horror over and over in their dreams.

Finnish psychologist and philosopher Antti Revonsuo has explained these and other observations with the idea that dreams could be a safe way to rehearse a dangerous scenario, just as a trainee pilot goes through the motions in a flight simulator. 'A dream-production mechanism that tends to select threatening waking events and simulate them over and over again in various combinations would have been valuable for the development and maintenance of threat-avoidance skills,' he wrote.

Mignot has another idea. In those vertebrates with a developed cortex, REM was a way to inject an element of unpredictability into the thinking brain, to shake up existing connections and forge new ones. 'It was selected, I believe, to increase creativity,' he says.

Most people will be familiar with the light-headed, lucid originality that comes in the quiet aftermath of a dream. Given the bonkers, non-sequiturial and highly inventive nature of the dreaming state it would be surprising if REM dreams did not result in new neural connections. There are also anecdotes of dreams that have resulted in Eureka-like discoveries, like the German chemist August Kekulé's 1862 daydream of a snake seizing its own tail that helped him solve the ring-like structure of benzene and other aromatic organic compounds.

'The most exquisite creations of poetic fancy, have been engendered under these circumstances, and conceptions suggested to the dreamy consciousness, which have paved the road to fame, and fortune,' wrote psychiatrist Lyttleton Forbes Winslow (now famous for his meddling role in the Jack-the-Ripper case) in his 1860 treatise *On Obscure Diseases of the Brain*. 'During the hours of sleep, the intellect has, with rapid facility, solved subtle questions, which had puzzled and perplexed the mind, when in full and unfettered exercise of its waking faculties. Difficult mathematical problems; knotty and disputed questions of science and morals; abstruse points of philosophy.'

Where non-REM is characterised by highly synchronised waves of electrical activity throughout the cortex, REM is far more random, says Mignot. 'This creates novel, more creative associations that help us to solve problems,' he suggests.

* * *

Philippe Mourrain, a developmental geneticist at Stanford, is hoping to find the secrets of sleep in a simpler creature than humans: zebrafish. When I ask him the same question I put to Dement and Mignot – 'What is the function of REM sleep?' – he immediately picks me up on the term REM. 'I prefer to call it paradoxical sleep,' he says, echoing fellow Frenchman Michel Jouvet. Fish don't have eyelids and they don't move their eyes when they're asleep, but they do experience a state similar to the one that Aserinsky and Kleitman described in humans, he says.

Hold on. If a fish can't close its eyes and there are no obvious rapid movements of its eyeballs, how do you know it experiences REM? As far as Mourrain is concerned, the main feature of REM or paradoxical sleep is the muscle paralysis that occurs all round the body rather than the peculiar twitching of the eyeballs that occur in humans and a few other mammals. 'Eye movement is not the best way to quantify this state,' he says.

Mourrain pulls up a slide on his laptop. It shows three schematic brains. At the bottom, there's a zebrafish brain, a slim, streamlined affair, with the bobbly olfactory bulb to the left and the spinal cord to the right. Mourrain points to the cortex, a tiny sliver of yellow sitting on top. The brain in the centre belongs to a mouse and its layout is strikingly similar. The most obvious difference is in the cortex, which looks like it's making a bid to engulf the other structures. At the top of the slide is the relatively recent evolutionary innovation that is the human brain, and here the cortex has mushroomed out of all proportion. The other structures of the brain are all present and correct, but like a fly trapped in amber they have been completely buried by a thick, cloven layer of yellow.

'We have a very human-centred definition of sleep,' says Mourrain. Looking at these three diagrams it's easy to see why. In humans, the primitive brain is completely inaccessible, so we've ended up identifying non-REM and REM by measuring the superficial waves of electricity that break across the surface of the cortex, the muscle tone and the mysterious movement of the eyes. If we define REM according to these strictly mammalian criteria, then it's difficult to capture this state in the zebrafish and other animals where the cortex is not developed. In fact, says Mourrain, all this means is that the current definition of non-REM and REM is simply inadequate. It would be much better to come up with a subcortical definition of sleep, he says, one rooted in the parts of the brain that actually control these phenomena.

In the zebrafish, these ancient structures are completely exposed. As a bonus, the larvae are transparent so the entire nervous system – brain and all – is visible under the low power of a dissecting microscope. With some clever genetic engineering, it's also possible to smuggle a jellyfish protein into highly specific populations of neurons, so that when they fire they emit a flash a fluorescent light. 'I'm now going to show you what REM looks like in a zebrafish,' says Mourrain.

A larval zebrafish fills the laptop screen, with the colour

saturation set so it appears blue on a black background. Certain parts of its brain are fluorescing, some more than others. There is a bright signal of neural activity coming from the hind-brain, a fainter glow in the vicinity of the eyes and a still-weaker, more diffuse emission from the muscles of the tail.

Mourrain hits the play button on the movie file. In the first few seconds, not much happens except for some almost imperceptible flickering of the olfactory bulb. The fish, with its head in a blob of transparent gel to keep it under the lens of the microscope, is awake. But when a drop of a hypnotic drug known to trigger REM is plipped into the water around the zebrafish, there is a neurological reaction that takes me back 30 years to a chemistry lesson in which I waved a strip of magnesium in a Bunsen's roaring flame. Boom! There is a burning flash of light, which starts in the fish's pons. A wave washes from the brainstem forwards through the brain, bleaching the eyes and petering out at the tip of the nose just like the waves that drive REM in mammals.

Mourrain replays the clip in order to draw my attention to something else. 'Concentrate on what happens to the focus of the image,' he says. As the pons flashes, the zebrafish blurs. 'It's going out of focus because of the muscle relaxation triggered by paradoxical sleep.' The beautifully coordinated wave of light radiating from the pons and accompanied by paralysis of muscles is exactly what occurs in mice, cats and humans during REM or paradoxical sleep. 'We can also see slow-wave-like sleep in these fish,' he says.

I ask if I can see some real fish and Mourrain is only too happy to oblige. In the basement of the building, he and his team look after about 20,000 fish. He swipes his way into one of several windowless labs, this one containing around 1,400 shoebox-sized tanks stacked on racks like books in a library. I step into one of the aisles to take a closer look at one of the tanks. The water is fed oxygen through a blue plastic tube and I count ten tiny zebrafish. There's a label on each tank. 'They all house different mutants and different transgenic lines,' says Mourrain.

On a work surface in the corner of the room, there's a small platform tilting in a circular fashion. On top of it, there are two objects wrapped in tinfoil, each about the size of a small bullet. 'It's a sleep deprivation experiment,' explains Mourrain, leaving me to picture the tiny zebrafish in the darkness of their little plastic tubes, the water around them swirling in constant, sleep-disturbing motion.

'If I could have a transparent human being, easy to manipulate and easy to image, then maybe I'd consider it as a model species,' says Mourrain. In the absence of this imaginary hominid, a creature like the zebrafish comes a pretty close second, and because of the relative simplicity and accessibility of its nervous system it might even be preferable. This is a fully transparent animal with the same kinds of basic neural circuitry and similar sleep states as in humans.

Mourrain's work on zebrafish strongly suggests that non-REM-like and REM-like states are both extremely ancient phenomena that arose more than 500 million years ago and have been conserved throughout the course of evolution. It's not beyond the realm of possibility that the REM-like sleep of fish could even trigger some kind of dream-like experience in their sliver of a cortex.

It's understandable, of course, that humans should be interested in human sleep. But trying to run before being able to walk is rarely a successful endeavour. 'Studies of non-mammalian vertebrates like fishes, but also amphibians, reptiles and birds, may bring more light than originally expected on mammalian sleep and REM,' says Mourrain.

# 5

# Sleeping dogs don't lie

*'Scientific research is one of the most exciting and rewarding of occupations. It is like a voyage of discovery into unknown lands, seeking not for new territory but for new knowledge.'*

Frederick Sanger

Bill Dement's talk was coming to a close. The year was 1972. As most people in the audience at the annual convention of the American Medical Association in San Francisco had never heard of narcolepsy, he'd taken along a video cassette of a patient in the midst of a cataplectic attack.

Dement was gathering his notes off the lectern after his presentation, when a vet called Benjamin Hart stepped up from the auditorium to say that one of his patients – a Doberman pinscher – had done something remarkably similar. Hart had diagnosed severe epilepsy and had the dog put down, which was a bit of a shame. But it gave Dement an idea. He remembers thinking: 'If there's one, there must be more.' If he could get hold of one, perhaps he could use it to identify new drugs that might help people with narcolepsy. He put out the word amongst the veterinary community across North America that he was on the lookout for narcoleptic dogs.

In April 1972, a toy poodle in Canada produced a litter of four. Eager families were quick to snap up the cute pups but one of them, a silver-grey female called Monique, soon developed what its owners described as 'drop attacks' when she tried to play. These were mostly partial paralyses, her hind legs going weak, her bottom slumping to the floor and her eyes becoming still and glass-like. At other times, particularly when fed, Monique would be struck by a full-blown attack. Something was clearly not quite right.

When vets at the University of Saskatchewan observed Monique, they suspected narcolepsy. As luck would have it, their diagnosis coincided with the arrival of Dement's peculiar circular, and they wrote back to him immediately. With Monique's owners persuaded to relinquish their pet, all that was needed was to figure out a way to get her to California.

Dement approached Western Airlines, the main carrier at the time, and instantly ran into a problem. They had a strict 'no sick dogs' policy. As the weeks turned to months, he began to entertain the idea of driving all the way from Stanford to Saskatchewan and back, a round-trip of over 3,000 miles. 'Under normal circumstances it would probably have been worth it,' Dement reflects, rocking back and forth in his squeaky chair. But president Richard Nixon had just weighed in on the Yom Kippur War on the side of Israel, causing the Arab members of OPEC to impose an oil embargo on the United States. 'There was a huge gasoline crisis' and 'a good chance of getting stranded,' he says. On my flight from London to Stanford, I had passed directly over Saskatchewan and I shivered at the thought of running out of petrol in this empty, icy wilderness.

'It's not a sick dog. It's a dog with a brain abnormality. It's an animal model of an important illness.' With some political lobbying, Dement succeeding in persuading Western Airlines to fly Monique to San Francisco, where she became something of a celebrity. 'Monique is very likely to collapse when she's eating something she especially likes, or when she smells a new flower outside, or romps around,' Dement's colleague Merrill Mitler told the Associated Press for a story that ran in dozens of newspapers across the US. There was a plan, he revealed, to get Monique to breed, with a view to creating a lineage of narcoleptic dogs. 'We hope to discover exactly where in the brain the dysfunction occurs that causes narcolepsy,' he said. 'This could be the first step towards developing a cure.'

As word began to spread, Dement and Mitler soon found themselves looking after Monique and three other narcoleptic poodles. For some reason though, none of their offspring (or indeed the offspring's offspring) showed even the merest hint of the condition. All was not lost. They had also acquired narcoleptic dogs from a host of other breeds, including more exotic individuals like a Chihuahua-terrier cross, a wirehaired griffon and a malamute. While none of these dogs seemed to pass narcolepsy to their pups, there were two breeds that did: Labrador retrievers and Doberman pinschers.

Dement remembers a breakthrough moment: a litter of around seven Doberman puppies. 'Several weeks after they were born, if you played with them they'd go down. Somewhere I've got pictures of them lying on the ground,' he says, gesturing around his shack-like office. I ask if I can see them. But Dement is not sure where they are. Looking at the reams of papers, books and memorabilia assembled on shelves, in drawers and over every horizontal surface, I can well imagine this is true. 'I'm retired and I don't have an assistant any more.'

Dement took the decision to lump for the pinschers rather than the retrievers, and by the end of the 1970s he was the proud custodian of a colony of over 30 narcoleptic Dobermans. It was clear that in this breed, at least, narcolepsy was inherited, caused by the transmission of a single, recessive gene.

How to find it? This task fell to Emmanuel Mignot, then a young, ambitious doctor just graduated from the University of Paris and recruited to Dement's lab in 1986.

\* \* \*

The phone on Mignot's desk rings and he answers. 'We have a delivery,' he says. I've come half way round the world to interview Mignot but I have also been looking forward to meeting Watson, his narcoleptic Chihuahua, and Mignot's wife has just dropped him off at reception.

I wander over to the floor-to-ceiling window that looks out onto a car park. On a sliver of wall beside it, there are two pictures, one a black-and-white photograph of a much younger Mignot, sitting cross-legged in a white coat with an imperious-looking Doberman at his side. Beneath the photo is a miniature poster, with a time-lapse sequence of photographs of a dog experiencing cataplexy.

Watson comes trotting in on Mignot's heels. Then the dog sees me and jumps backwards, instantly on his guard. I get down on the floor in an effort to put him at ease, but this provokes a sharp yap and he springs in at me, then out, as if sparring for a fight. I offer my hand and he approaches slowly this time, close enough for me to stroke him.

Watson is not the first narcoleptic dog that Mignot has adopted. Before him there was Bear, a sprightly Belgian schipperke with narcolepsy. He was 'a wonderful dog' and 'no trouble at all', says Mignot, except perhaps for the fact that his cataplexy had a rather inconvenient trigger. 'He would fall over when he was having a crap.' I have watched a YouTube video of Bear posted by Mignot's wife in 2010 and he was a very fluffy dog. I imagine these cataplectic fits could have got messy. 'Yes they did,' Mignot acknowledges.

When Bear died of old age in 2014, Mignot did not imagine that he would adopt another narcoleptic dog. Then along came this Chihuahua. 'It's such a silly breed,' says Mignot. 'Not one I would ever have chosen myself.' But people with narcoleptic dogs keep on offering them to the Center, and Mignot thought 'why not?'

Mignot has seen hundreds of people with narcolepsy during his medical career and has published almost as many research papers on the condition, so he understands his patients better than most. But I guess these two dogs – Bear and now Watson – have given him a still greater insight into what it's like to live with narcolepsy. 'Absolutely,' he says.

If the Chihuahua is called Watson, I wonder if Mignot sees himself as molecular biology's Sherlock. The idea appears not to have occurred to him, yet locating the gene responsible for canine narcolepsy required an extraordinary level of sleuthing, a discovery that revealed one of the most fascinating neurological pathways in the brain, a complex network involved in the regulation of wakefulness and sleep and much more besides. I silently resolve that when I get back to London I will mail Mignot a deerstalker.

\* \* \*

Looking back to the 1980s and his ambition to find the gene responsible for canine narcolepsy, Mignot admits to being 'a little naïve'. Breeding narcoleptic Dobermans is harder than it sounds, as the afflicted tend to topple over while mating, temporarily paralysed by a cataplectic thrill. This practicality aside, there was also the titanic task of locating a gene whose sequence was not known in a genome that was, at the time, a no-man's land. 'Most people said I was crazy,' says Mignot.

In some sense, his critics were right, because it took Mignot more than a decade, hundreds of dogs and over one million dollars. He was nearly beaten to it too.

In January 1998, just as Mignot's mapping team was closing in on the gene, Luis de Lecea and his colleagues at the Scripps Research Institute on the outskirts of San Diego published a paper describing two novel brain peptides. They gave them the name 'hypocretins' – an elision of hypothalamus (where they were found) and secretin (a gut hormone with a similar structure). They appeared to be chemical messengers acting exclusively inside the brain.

Just weeks later, a team led by Masasi Yanagisawa at the University of Texas independently described the exact same peptides, though they called them 'orexins' and identified the structure of the receptors into the bargain. They speculated that the interaction of these proteins with their receptors might have something to do with regulating hunger or feeding behaviour. There was no mention of wakefulness. No mention of sleep.

Back at Stanford, Mignot and his colleagues heard about the hypocretin and orexin papers. By then, they had the mutation responsible for canine narcolepsy narrowed down to 50 genes, and among them was one of the hypocretin receptor genes. 'We should look at that gene,' suggested then-postdoc Ling Lin. 'It's in the brain, in the hypothalamus, maybe it's involved in narcolepsy,' she said. The ever-methodical Mignot insisted on reducing the number of candidate genes still further before looking at any of them in detail. 'If I had followed Ling's advice, we would have found the mutation a year earlier.'

By spring of 1999, Mignot and his team had the shortlist

down to just two genes. One was expressed in the foreskin. 'It didn't look like a candidate for narcolepsy,' he admits. The other, encoding one of the two hypocretin receptors, was still in the running. It was looking increasingly likely that the hypocretin system might somehow be involved in narcolepsy.

That June, at the annual meeting of the American Academy of Sleep Medicine, Mignot was keeping his suspicions to himself, when a sleep scientist he'd never met cornered him and began to talk of a discovery in his own lab, claiming he had a mouse model of narcolepsy. 'I see that pretty much every meeting,' says Mignot, 'so I rarely get excited.' But as he began to extricate himself from the conversation, his eye caught the scientist's name badge: Christopher Sinton, University of Texas, the precise same institution as Yanagisawa, who had described the hypocretins (or orexins, as he called them) the previous year.

Mignot's mind began to race. If Yanagisawa had figured out that the hypocretins had a role in the regulation of sleep, the obvious thing to do would be to team up with a sleep researcher at the same university. 'We have this mouse and we think it has narcolepsy,' said Sinton. 'It seems to go directly into REM sleep.'

Yanagisawa and his colleagues must have engineered a mouse without hypocretins or hypocretin receptors, thought Mignot. 'My heart was racing at about 250 beats per minute.' Sinton showed a few EEG recordings from the mouse, revealing all the characteristics of narcolepsy and began to ask about the progress in mapping the dog narcolepsy gene. Mignot contained his excitement and kept schtum. Sinton made no mention of the hypocretins or Yanagisawa, but as soon as Mignot got away he called his lab. 'Work like donkeys,' he told them. 'There's a possibility that someone else is on the same trail.'

In Mignot's office at Stanford, I notice a framed photograph propped up at the back of a bookcase. It shows fuzzy bands of DNA fragments separated according to their size. There is a clear outlier, a lane in which a chunk of DNA from a narcoleptic Doberman has not made it nearly as far as the others. It contains the instructions to build one of the two hypocretin receptors, but its electrophoretic journey has been hindered by a huge and apparently random sequence of DNA lodged right in the middle of the gene. 'It was incredibly exciting,' says Mignot.

Within weeks Mignot and his team had a paper ready for publication, and it appeared in the prestigious journal *Cell* that August. Kahlua, one of a litter of Dobermans named after alcoholic drinks, made it onto the cover along with the tagline 'Gene for Narcolepsy'. Inside, Mignot and colleagues described multiple mutations responsible for canine narcolepsy in exacting details. The implications for human narcolepsy were clear: 'Abnormalities in the hypocretin neurotransmission system are thus likely to be also involved in human cases,' he wrote.

Yanagisawa and colleagues added their experimental evidence to the mix just two weeks later, also in the journal *Cell*. As Mignot had suspected from his conversation with Sinton, they had engineered a knockout mouse, one unable to make any hypocretins. Lo and behold, these deficient rodents were 'strikingly similar to human narcolepsy patients', suffering sleep attacks many times an hour when they would normally be active.

When Bill Dement succeeded in getting Monique flown from Saskatchewan to San Francisco in 1973, DNA sequencing was a technology that had not been invented. He could not have possibly known where the sleeping dogs would lead. It's a story of the serendipity of science and of the benefits of investing in research for the long-haul. 'There were several times when the administrators wanted to kill the dog colony,' says Mignot. But he knew that he'd get there in the end.

With the Dobermans revealing that narcolepsy appears to be a result of a malfunctioning hypocretin system, their job was done. In 1999, Dement and Mignot wound up the colony and sought homes for all the dogs in their care. The rate of canine narcolepsy was higher around Stanford than in any other city in the world for quite a while.

Though the dogs had gone, the work had only just begun.

\* \* \*

Under normal circumstances, a neurotransmitter and its receptor work a lot like a key and lock. A key (the neurotransmitter) fits into a lock (its receptor) to open a door (effect a change within the target neuron). In the case of Mignot's Dobermans, a piece of chewing gum (a massive mutation) had jammed the lock (the hypocretin receptor), rendering the key (the hypocretin) completely useless. Whether it's the lock that doesn't work (as in the case of Mignot's Dobermans) or the keys are missing (as they were in Yanagisawa's knockouts), the upshot is the same. The door won't open. The hypocretin system is broken.

I have spent many, many hours trying to make sense of the hypocretin pathway and render it faithfully on the page. Circuits, by definition, have no beginnings and no ends and breaking into one in an effort to provide a linear account of cause and effect fails to capture the significance of the looping circuitry.

In the paper that first described the hypocretins back in 1998, it transpired they were being expressed exclusively within the brain, produced in the cell bodies of neurons occupying two tiny spots within the hypothalamus, one on either side of an imaginary vertical slice through the centre of the brain. This much I get.

But where do these neurons project to and what might the hypocretins be up to? When the scientists followed up by tracing the path of these neurons throughout the brain, they found them projecting to the locus coeruleus, the septal nuclei, the bed nucleus of the stria terminalis, the paraventricular and reuniens nuclei of the thalamus, the zona incerta, the amygdala and on and on, a list of Greek and Latin destinations so extensive it's like being trapped in a labyrinth. If I am to make sense of all this, I need to talk to Luis de Lecea, a man who quite possibly knows more about the hypocretins than anyone else alive.

* * *

Back in 1998, when de Lecea and his colleagues first described these then-mysterious peptides and called them hypocretins, he was in his mid-20s, and had only recently moved from Barcelona in Spain to San Diego. In 2005, he made the move to Stanford University, which is where I catch up with him in the hope that he can bring me up to speed on the latest thinking about the hypocretins. 'I thought we'd have made more progress,' he admits early on. 'I thought we'd understand everything about them.' The fact that there is still a lot to discover about the hypocretins is testimony to just how multi-talented these neurochemicals are.

When a nerve cell fires in the brain, an electrical signal ripples along the membrane in much the same way that an electric current passes along a wire. When the pulse reaches the end of the nerve, the synapse, neurotransmitters erupt from the inside to the outside of the cell, chemical messengers that drift across a small void or synaptic cleft to a target neuron. There they bind to highly specific receptors on its surface, and in so doing make it more or less likely to fire.

In addition to these rather direct cellular go-betweens, there are also neuromodulators, chemicals that act a little more like hormones than textbook neurotransmitters. Once they are released into the space between neurons, they tend to float further afield, influencing the activity of many different cells throughout the brain in many different ways. In some settings, the hypocretins act as neurotransmitters. In others, they are neuromodulators. With these basics established, de Lecea outlines three major roles for these powerful molecules.

The first and most important of these is to send a wake-up signal to the cortex, the thinking part of the brain. This is achieved in a relatively simple manner. When a hypocretin neuron fires, it's a little like a smouldering taper setting off a firework. As the rocket launches into a pitch-black sky, so a nervous impulse shoots from the hypothalamus down the brain stem to a node called the locus coeruleus. The firework erupts, sending stars in all directions, and the locus coeruleus emits a volley of messages to every corner of the brain, the excitatory neurotransmitter noradrenaline switching on neuron after neuron after neuron.

If the hypocretin neurons keep on firing, it's like sending up rocket after rocket, so fast that the sky remains ablaze with light. As I follow through this analogy, I realise that having narcolepsy is like turning up to a fireworks display to find there are no rockets. A few kids are waving sparklers, someone lights a Roman candle and a Catherine wheel inspires a half-hearted 'ooh' and a reluctant 'aaah' before spinning off into the bushes. There's no getting round it. The party sucks.

There's lots of research that confirms this rocket-like role for hypocretins. Easily the strongest evidence comes from opto-genetics, an incredibly powerful approach for studying the brain that de Lecea had a hand in pioneering.

'If you shine a light in the brain nothing happens,' says de Lecea. But with a bit of molecular cunning, involving a virus, a promoter and a gene found in blue-green algae, it is possible to transform a particular population of neurons so that it becomes sensitive to light. 'You can activate them, inhibit them, depolarise them with just a pulse of light.' It sounds miraculous. Anticipating my wish to see it in action, he swivels to his computer screen and brings up a PowerPoint presentation.

When I was at university, I remember hearing about the work of Canadian-American neurosurgeon Wilder Penfield. In the 1930s, he devised a radical method for treating severe epilepsy, keeping his patients conscious during brain surgery. By monitoring their reaction to delicate prodding with an electrode, he was able to locate and remove the region causing seizures. In so doing, Penfield discovered that specific regions of the brain appear to be associated with very specific sensations or memories, this region triggering a tingling in the left thumb, that area stimulating the sound of an orchestra, another spot conjuring up a childhood memory of being by a river.

In 2005, researchers working at the Montreal Neurological Institute at McGill University in Canada (the same institution where Penfield had carried out this work) attempted some-thing similar, only using an incredibly fine glass pipette to record the activity of individual hypocretin cells deep within the brains of mice. They found these neurons firing during wakefulness and silent during sleep.

When de Lecea read the paper he was excited, but he could

see it had obvious limitations. Sticking a pipette into a brain and hoping its tip hits home is a bit like shoving an arm into a haystack and hoping to pull out the proverbial needle. What's more, while the pipette method allows measurement of cell activity, it doesn't give a reliable way to take over control of the cell, to manipulate its activity at whim. This is where opto-genetics comes in.

De Lecea sets a video to play. There is a mouse in a cage with a thin fibre-optic cable running into its brain. 'The mouse is asleep,' he says, crests of slow, non-REM sleep spooling across an inset video at the top of the screen. 'In a moment, you'll see the light travelling down the fibre-optic cable.' A second passes. Two. Three. Then, just as I begin to think that it must be the wrong movie, the optic cable comes alive, a pulse of bluish light flashing for precisely ten seconds.

Nothing happens.

I know that nerve impulses zip along at tens of metres a second and usually have an almost instant effect. I look at de Lecea for an explanation of the failure of the light to do any-thing at all. 'Wait,' he says, with a flicker of a smile, and I turn back to the screen.

Almost 20 seconds after the flashing started something remarkable happens. The mouse jolts awake. I can almost feel the hypocretins working their magic, igniting the locus coer-uleus and lighting up the mouse's brain. 'It takes almost 20 seconds to wake up, so there's some computation going on,' says de Lecea. The rodent looks around, its whiskers twitch-ing. Then, because ten seconds have passed and the light has gone off, it falls asleep as suddenly as it awoke. There can be few more striking illustrations of the power of the hypo-cretins – the neurochemicals I've lost – than this. Completely unexpectedly, I feel my tear ducts tingling and, for a fleeting, still-speechless second in which I ignore the fibre-optic cable running into its brain, I actually envy the mouse.

De Lecea moves on to outline a second major function of hypocretin cells. In addition to rocketing to the locus coeruleus, these special neurons also project to regions of the brain that signal with the neurotransmitter dopamine. In this context, the hypocretins play the role of neuromodulator, helping target cells to make sense of all their other inputs and produce the appropriate response at the appropriate moment. Without hypocretins, the dopamine cells lose their focus. The release of dopamine is 'a little more diffuse,' says de Lecea.

Dopamine is essential for the processing of reward, in planning and for motivation. A pleasant experience, like eating a square of chocolate perhaps, is likely to increase the levels of dopamine in the brain, upping the motivation to have a second chunk, or maybe even a cheeky third. Indeed, addictive drugs like nicotine, cocaine and amphetamines act directly on the dopamine system, resulting in an almost irresistible motivation to seek out another fix.

The link between the hypocretins and dopamine makes sense of a feeling that people with narcolepsy often report, that they struggle with planning and motivation. In a brain where the circadian rhythm brings about a healthy surge of hypocretins as morning approaches, there's likely to be a neat and tidy release of dopamine and a strong motivation to clamber out of bed and head for breakfast. In a narcoleptic brain, should it wake up at all, the dopamine signal is likely to be fuzzy and there will be less get up and go.

The dopamine story has also led to the intriguing idea that people with narcolepsy should be less prone to addiction. When scientists push hardcore drugs like cocaine and heroin on laboratory animals, it doesn't take long for them to show signs of addiction. If the animals are narcoleptic, however, the craving is a little less. 'It's really subtle,' says de Lecea. This doesn't mean I could develop a cocaine habit with no comebacks. But it does mean that if I subsequently attempted to clean up my act, I might find it easier than an addict with a fully functioning hypocretin system. 'It would reduce the cravings,' he says. In theory then, a drug that messed with the hypocretin system might be helpful for breaking an addiction.

The hypocretins also affect mood, through the aforementioned pathways but also through the modulation of neurons that produce serotonin, a neurotransmitter strongly associated with depression. 'Without hypocretins, the fluctuations of mood are going to be less pronounced,' says de Lecea.

There is a fourth and exceedingly important function that hypocretins perform: they exert a tight control over the region in the pons that Michel Jouvet recognised as the trigger for REM. This explains why many people with narcolepsy (hence no hypocretins) tend to trip into REM so soon after the onset of sleep. It is also likely to account for the cataplexy. Without hypocretins, the amygdala overdrive in response to emotions goes unchecked and seems to activate the muscle atonia pathway that is usually only brought into play during REM.

With the hypocretins affecting at least three major neurological pathways and many other networks that we are only just beginning to understand, it is easy to see why they are so important. It also explains why people with narcolepsy can get a little obsessive about these neuropeptides. The discovery that this system is defective in sleeping dogs has given weight to what people with narcolepsy already knew, that there had to be something seriously wrong with their brains. The awareness of this pathology has made the narcoleptic journey a little easier, offering validation for patients, forcing doctors to acknowledge the condition, facilitating diagnosis and reducing the risk of being dismissed as a basket case along the way. Few sleep disorders have this kind of clarity.

Even more importantly, it could lead to far more effective medication. Up until now, the only way to treat narcolepsy has been to manage its symptoms, mainly through a cocktail of drugs. Most of these, if they work at all, are usually only partially effective, like putting a sticking plaster on a gaping wound. This is because they were, for the most part, brought into play before anyone knew of the existence of hypocretins. Nobody really knew how these drugs were helping, but somehow they were. With the discovery of the hypocretins and the more-than-20-year research to figure out what they do and why their absence might result in the symptoms of narcolepsy, it should now be possible for pharmaceutical companies to develop drugs that target the hypocretin system with far greater precision than is currently possible (of which more later).

The significance of the hypocretins doesn't stop there. Everyone needs to know about these two nifty peptides. 'There are thousands of publications on the hypocretin system and it's not because it has more projections than others,' says de Lecea. 'In most other neural networks, there are parallel and multiple layers of security,' so if something isn't working properly there are systems that can step in and pick up the slack. In the case of the hypocretins, however, there appears to be no backup at all.

This is not particularly good news for people with narcolepsy (though it does explain why losing hypocretins should be such a big deal). But it does mean that manipulations of this system produce the kind of clear-cut response that scientists can work with. 'It is a brilliant model for understanding neural networks more generally,' says de Lecea. The hypocretins and the many surprising things they do are giving us a privileged insight into how the human brain does what it does.

# 6

# Bad breath

*'Laugh and the world laughs with you,
snore and you sleep alone.'*

Anthony Burgess

I blame the comics.

If a character in a comic strip or a cartoon sequence is asleep, more often than not they are snoring, a long line of Zs lazily drifting into the air. As a result, most of us have grown up to think of snoring as part and parcel of sleep.

In Walt Disney's *Snow White and the Seven Dwarfs*, we smile at the endearing range of noises that Sleepy, Doc, Happy, Sneezy, Dopey and Bashful emit as they kip. Grumpy is disturbed by the racket, but we don't for a minute entertain the possibility that all the noise might be the cause of his moodiness.

Indeed, if we respond to snoring at all, it's usually to laugh. Few of us see snoring as a sleep disorder, let alone one that can have serious, even life-threatening consequences. Purrings, occasional gurgles and gentle buzzes during sleep might not seem like too much of a concern, but when these become persistent snorts, whistles and stentorian raspings, snoring is harder to ignore, particularly if you are the other person in bed.

But beyond the devastating impact that persistent snoring can have on relationships, it's dangerous too. The data on snoring are noisy. In some sleep labs snoring is not measured at all. In others, the clinician will scribble down a single, subjective estimate of its severity. In others, a microphone will record snoring episodes and their severity and map them onto the architecture of the patient's sleep. So the prevalence of snoring depends on who you listen to. In some studies, just 1 in 20 adult men and 1 in 50 adult women are snorers. In another, where the snoring box is obviously more readily ticked, more than 8 in 10 men and 5 in 10 women snore.

Whatever the true prevalence of snoring, it lies at the thin end of a big, fat wedge, just one of many forms of what sleep clinicians refer to as 'sleep disordered breathing'. All of them, including snoring, obesity hypoventilation syndrome, central sleep apnea, obstructive sleep apnea and upper airway resistance syndrome, can have serious consequences for health.

* * *

Charles Dickens conjured up some fabulous characters but rarely out of thin air. He based many of them on real people, minutely observed. He was very clued up about medicine too, and visited genuine diseases, disorders and conditions on many of his characters. Several of them seem to suffer from bona fide sleep disorders, but nowhere is this clearer than in the case of the servant Joe, 'a fat and red-faced boy' who makes his comedic appearance in *The Pickwick Papers*.

Joe first pops up at a military review in Rochester, an event at which the soldiers are to act out the storming of a mock citadel for the edification and entertainment of the public. Joe's job – serving the needs of his master Mr Wardle and friends – is rather complicated by the ease with which he drifts out of consciousness. 'Joe! – damn that boy, he's gone to sleep again,' says Wardle on several occasions. Joe even sleeps through much of the display, 'as soundly as if the roaring of cannon were his ordinary lullaby'. He is 'a natural curiosity,' Wardle boasts. 'Goes on errands fast asleep, and snores as he waits at table.'

Dickens' portrayal of medical conditions was so believable that doctors often used them as reference points for their own diagnoses. Take the case of PS, for instance, a short, fat gentleman who ran a poultry business in nineteenth-century Liverpool. 'The tendency to sleep became constant and overpowering', and had infected all aspects of PS's life, his doctor Richard Caton reported to his medical colleagues at the Clinical Society of London.

'He had been fond of attending the theatre, but now he slept soundly through the most exciting drama.' On one occasion, when PS had had guests over to dinner, he'd fallen 'soundly asleep whilst carving the joint at his own table'. Sitting in the audience, Christopher Heath, the society's president, immediately thought of 'the classical case of the fat boy in Pickwick'.

When out on the town, PS could fall into a deep sleep while standing or even walking. He'd frequently collide with other pedestrians, sometimes with lampposts. Open-topped buses proved an unexpected hazard. 'He repeatedly fell fast asleep while reading on the top of an omnibus, and had many narrow escapes, fellow passengers several times seized him when on the point of falling head first from the roof.'

The sleep hit PS's poultry business too. In his shop 'he would waken and find himself holding in his hand the duck or chicken which he had been selling to a customer a quarter of an hour before, the customer having meantime departed'. Within a few years of becoming overly sleepy, his business had gone bust.

More than a century after the publication of *The Pickwick Papers*, a business executive presented similar symptoms to his doctor. He'd been struggling to stay awake during the day for some time, but it was a card game that had finally prompted him to seek medical help. He'd been dealt an almost unbeatable hand, a full house comprised of three aces and two kings. Sadly, however, he'd dropped off to sleep and failed to capitalise on his luck. A few days later he checked in to the Brigham Hospital in Boston.

At a little over five foot and weighing almost 20 stone, his 'general contour was strikingly similar' to the cartoon-like illustrations of Joe the fat boy. The doctors soon discovered the problem. At night, the businessman's breathing was slow and shallow, probably because of his obesity. His brain just wasn't getting the oxygen it craved and carbon dioxide was building up in his bloodstream.

Most of us are equipped to respond to this situation. As carbon dioxide accumulates in the blood, some of it dissolves in the water to form carbonic acid, causing a drop in pH that is picked up by litmus-like chemoreceptors in the blood vessels and at the base of the brain. These receptors respond by instructing the muscles to the lungs to contract faster and more strongly, rapidly restoring the balance of oxygen and carbon dioxide. In the case of the poker-playing businessman, however, he was fighting a losing battle to ventilate his lungs and living in a state of near-permanent acidosis. The chemical alarm signals were there, but his chemoreceptors had simply stopped responding, just as a parent can end up ignoring the repeated sleeve tugging of a child. It's not entirely clear why this should result in daytime sleepiness, but acidosis could have a general anaesthetising effect on neural activity, thereby increasing somnolence.

Based on the superficial similarities with Joe the fat boy, the Brigham medics coined the term 'The Pickwickian Syndrome' to describe the poker-playing businessman's symptoms. It was one of the first clear indications that difficulties with breathing can have very serious consequences for sleep and wakefulness. The Pickwickian Syndrome (or obesity hypoventilation syndrome as it's more properly known) remains pretty rare, but it paved the way for the discovery of a related condition in which the breathing isn't merely laboured. It stops completely.

The first formal description of someone whose breathing actually stopped during sleep was of a man called Etienne. Serving in the French army during World War II, he'd climbed a tree to escape 'a barrage of bullets and other missiles', using his belt to strap himself to a branch. With the battle raging around him, he'd promptly fallen asleep. When he eventually woke up, everything had gone quiet. Disorientated, he climbed down and was captured by the Germans. Eventually, at a sleep clinic in Marseilles in the 1960s, doctors found that although his diaphragm and lung muscles continued to contract during sleep, there were frequent periods when no air passed his lips for around 30 seconds; then he'd suddenly gasp, his breathing would resume and he'd sink back into a deeper state of sleep. These 'episodes of obstructive apnoea' were messing with his sleep cycle. 'Our patient's night sleep is greatly reduced,' wrote the neurologist Henri Gastaut and his fellow medics, resulting in bouts of uncontrollable sleepiness during the day.

Even then, it was another decade before the medical profession really began to cotton on to the extent of this problem.

* * *

In the early 1970s, at the Royal Victoria Hospital in Montreal, a young medic called Meir Kryger was caring for a man with a mysterious constellation of symptoms – obesity, type 2 diabetes, epilepsy-like seizures and excessive sleepiness – and nobody could figure out quite what was going on.

The case of Etienne over in France was not yet widely known, so when, one night, Kryger was doing the rounds and witnessed his patient's breathing come to a complete standstill it came as a complete surprise. 'Up to that point I'd never considered sleep to be a time of danger,' he says. In collaboration with a colleague from the Montreal Neurological Institute across the road, Kryger set about monitoring his patient's brainwaves and breathing during sleep. 'I thought maybe this was somehow going to explain the fact that he had all these other symptoms during the daytime,' he says.

Like Gastaut had done for Etienne some five years earlier, they documented clear instances where the man actually stopped breathing. In addition 'his heart really slowed down and actually stopped at times when he was struggling to breathe', sometimes for as long as ten seconds at a time. As with Etienne, the read-out from Kryger's patient revealed that each event resulted in 'partial arousal'. His punctuated breathing was puncturing his sleep.

When Kryger wrote up his investigation, he suggested a name for the condition: 'the sleep deprivation syndrome of the obese patient'. It's a good job the name didn't stick, in part because it's such a mouthful but also because this is far too specific. This phenomenon is not confined to the obese. In fact, it can occur in anyone and, as we will see, at any age.

It fell to another sleep specialist, Christian Guilleminault, to put 'sleep apnea' on the medical map in the mid-1970s. As a bonus, he devised a way to quantify its severity. The apnea-hypopnea index is a bit of a mouthful but it's simply the number of apnea events per hour. For adults, up to five events per hour is considered 'normal', 5–14 is 'mild sleep apnea', 15–29 is a 'moderate' case and 30 or more is 'severe'.

* * *

Guilleminault is a gentle man with a sparkle in his eyes. He leads me from the reception of the Stanford Sleep Center past several rooms used to conduct overnight sleep studies. He shows me briefly round Room No. 14, specially fitted out with a reinforced bed, a hoist and a bariatric toilet capable of accommodating the buttocks of a patient weighing up to 300 kilos. 'It's a relief we are not in the Midwest or we'd see a lot more patients like this,' mutters Guilleminault as the tour continues in the direction of his office.

It was early in his medical career that Guilleminault read about Michel Jouvet's groundbreaking work on animals showing that sleep is regulated in structures buried deep within the brain. He went to his boss at La Salpêtrière in Paris and made a case for starting a sleep clinic. 'He was absolutely not impressed by that idea,' recalls Guilleminault. 'He said: "Sleep is for dreamers."'

But confident that he was onto something, Guilleminault was persistent. Alongside his regular commitments, he secured a small grant, took a significant pay cut and co-opted the assistance of interns and students to establish what might be referred to as the world's first sleep clinic. It was a hard year, with Guilleminault carrying out his normal day job, then watching patients sleep before handing over to a junior colleague in the early hours of the morning so as to snatch a few hours' rest for himself. Over 12 months, from mid-1970 to mid-1971, he saw around 450 patients with a wide range of sleeping problems, including sleep apnea.

By November 1971, Bill Dement over in Stanford had persuaded Guilleminault to take a two-year leave of absence from La Salpêtrière and move to California. One of the first patients he saw in the US was Raymond, an 11-year-old boy who appeared in the pediatric intensive care unit shortly after Guilleminault had arrived in Stanford. The boy presented with off-the-scale blood pressure, which was unusual in such a young child. Nothing the doctors tried could get it down. The boy also snored and appeared to be very sleepy during the day, so the head of the ICU, out of ideas, asked Guilleminault to take a look. When he hooked up Raymond to the machines, he discovered that his breathing stopped over and over throughout the night.

Guilleminault scribbled his findings in the boy's notes, which caused a stir. 'In the US, medical notes are legal documents so there was suddenly an obligation to do something about what I'd seen.' But few people in the US had noticed this strange, stop-start breathing before, and the ICU team had no idea what to do about it. So they summoned this young French upstart Guilleminault to a formal meeting to explain. In a big amphitheatre, surrounded by stern, senior clinicians, he recounted everything that he knew from his experience in Europe.

When Raymond's consultant asked if there was a way to treat this breathing abnormality, Guilleminault informed him that the answer was a tracheostomy. 'You do a little hole in the neck,' he tells me, pointing an ET-like finger at a spot just beneath his Adam's apple.

The consultant then asked if Guilleminault had done a lot of these operations.

'No, Sir,' he replied.

'Do you mean to say this would be the first time anyone had performed a tracheostomy for this problem in a child?' asked the senior clinician.

'Yes.'

There was silence. The rather radical treatment being proposed had only ever been performed on a handful of adults far away in Europe. But there was an even bigger issue at stake.

'Do you know the colour of this child's skin?' Raymond was black. 'Are you really suggesting that we experiment on a black kid?'

With race an ever-sensitive issue across the US, Raymond's operation was out of the question. Which is how a white girl called Carlene, who presented with similar symptoms a few weeks later, became the first ever child to undergo this procedure for this condition. Like Raymond, she had high blood pressure, stopped breathing during the night and slept so much during the day that she'd been kicked out of several schools. Yet within three weeks of the operation, she had become an active child once more, albeit one with a small hole in her throat. With Carlene proving a successful guinea pig, Raymond was wheeled into the operating theatre, and his recovery was similarly spectacular.

Raymond and Carlene underscored something that Guilleminault already knew from his experience in Paris. Problems with breathing during sleep were not confined to the Pickwickian Syndrome, as most sleep specialists in Europe assumed at the time. Raymond and Carlene also confirmed that this was not a problem confined to the obese population, but could also occur in those of healthy weight and in children. It began to look as though sleep apnea might be far more common than anyone had hitherto imagined.

\* \* \*

Sleep apnea comes in one of two main flavours, though neither of them is particularly palatable. 'Central sleep apneas' originate in the respiratory centre in the brainstem, a cluster of neurons that integrates chemical, neural and hormonal inputs to generate an appropriate rhythm of inhalation and exhalation. When this centre takes a 'time out', the muscles responsible for the rise and fall of the chest and for tugging downwards on the lungs cease their contractions and breathing stops.

The other form of sleep apnea – obstructive sleep apnea – is far more common. When we fall asleep, most of the muscles around the body relax, including those at the back of the mouth. This would not normally be a problem, but it becomes one if the fleshy tissue in the throat closes in on itself in such a way that it blocks the airway. The respiratory centre continues to deliver signals to the breathing muscles, and these continue to contract, but they can do nothing if the passage to the lungs has closed. It would be like trying to open a set of bellows when the inlet is completely blocked.

Obstructive sleep apnea is thought to affect around one in 20 men and one in 50 women. But because it's hard to spot, there are surely lots of cases that are slipping under the medical radar. 'It's probably around twice this common,' says Kryger. 'We were thinking that this has got to be a very rare syndrome,' he remembers of those early days in the 1970s. But, in the space of a decade, sleep apnea had emerged as a major sleep disorder. 'Suddenly, everybody knew of someone with the condition,' he says.

Of course, sleep apnea is nothing new. Indeed, Kryger has found several plausible cases from antiquity. In Ancient Greece in the fourth century BC, for instance, a chap called Dionysius presented with such an 'extraordinary degree of Corpulency and Fatness' that 'he had much adoe to take breath' and sounded as though he had 'a fat hog lying upon the snout'. At the instruction of his doctors, Dionysius had his servants armed with needles – 'very long and small' – to be deployed in the event that he stopped breathing as he slept. Around 100 years later, the King of Cyrene (a Greek colony on the coast of North Africa) is said to have spent the last days of his life so 'weighted down with monstrous masses of flesh' that 'he choked himself to death'.

There are several factors that might explain why it took so long for anyone to notice sleep apnea. For a start, there's the old problem that sleep and its many different disorders are not considered a particularly high medical priority, even now. In essence, nobody was looking. It's also possible that sleep apnea was less common than it is today, with the rise in obesity in western society driving a relatively recent explosion in the prevalence of this disorder. In addition, the key pathology – the interruptions in breathing – usually go unnoticed because the individual is asleep. Most people with sleep apnea aren't aware of the partial arousals; if they wake at all, it's only for a second or two.

The frequent disruptions to the sleep cycle mean that someone with apnea rarely, if ever, reaches the deeper, more restful stages of sleep. There are also some very serious knock-on consequences for the cardiovascular system.

When the lungs are operating optimally, they facilitate an exchange of gases, the delivery of oxygen from the air to the bloodstream and the simultaneous removal of carbon dioxide from the bloodstream to the air. This gaseous transaction takes place between little air sacs called alveoli and a network of blood-filled capillaries that embraces them. A pair of human lungs typically boasts around 500 million alveoli, which if flattened out would cover a cricket wicket from stump to stump. The capillaries are spread over most of this area, giving plenty of opportunity for oxygen to come in and carbon dioxide to go out.

If the area for this exchange is reduced, it can put immense strain on the heart and blood vessels. There are lots of ways this can occur, but hypoxic pulmonary vasoconstriction is a pretty neat example, a crafty little trick we – and other animals – have evolved. If the oxygen supply to a part of the lungs falls, the capillaries to that region constrict accordingly, rerouting the blood to where there is greater promise of oxygen. This is rather like switching off a radiator in a room that's not in use, diverting heat to the parts of the house that are. It's the most efficient way to run your home.

The trouble is that if we breathe badly when we sleep, the oxygen levels can get so low and the blood vessels can constrict to such an extent that the blood starts backing up. In the central heating analogy, severe sleep apnea would be like keeping the boiler pump running at full whack but shutting down all the radiators. The pressure in the system would shoot up and either the pump would burn out or the pipes would explode. A central heating system, of course, is fitted with a pressure release valve to prevent such domestic disaster. The circulatory system is not.

The evidence that bad breathing during sleep has consequences for the heart and blood vessels is overwhelming. It's so closely associated with high blood pressure, heart flutters and an increase in heart muscle that it's hard to imagine this isn't cause and effect. It's unsurprising therefore that sleep apnea should increase the risk of cardiovascular disease.

The Wisconsin Sleep Study Cohort illustrates this pretty

convincingly, a study established in 1988 to try to get a handle on the causes and consequences of having sleep-disordered breathing. When obvious confounding variables like age, sex, body mass index and smoking are controlled for, the greater the apnea-hypopnea score the higher the incidence of coronary heart disease or heart failure. Those in the study with severe sleep apnea were also many times more likely to have died than those with no sleep apnea.

Consider this. For women, the prevalence of sleep apnea increases with age. The same goes for men but only up to a point. In fact, the proportion of men with sleep apnea reaches something of a plateau at the age of around 65. At first glance, this sounds like great news, like elderly men are somehow getting over their sleep apnea. The stark reality is that the proportion of elderly men with sleep apnea plateaus because middle-aged men with sleep apnea are dying.

Academics tend to be careful not to overstate their case, but I do wonder if their laudable efforts to acknowledge the complexity of cardiovascular disease have had the unintended consequence of understating the damage that bad breathing can do to the circulatory system. If someone survives a heart attack or a stroke, they will tell their family and friends that they had a heart attack or stroke. If someone suffers a cardiac arrest, the cause of death will be given as cardiac arrest. But how often is it that cardiovascular events like these are actually a direct consequence of sleep-disordered breathing?

The consequences of sleep apnea do not stop with cardiovascular disease. There is some evidence to suggest that apnea may be responsible for many cases of sudden infant death syndrome (SIDS), or 'cot death'. In the 1960s, an Australian doctor reported on three cases of SIDS. Each baby had been rushed to hospital after its alarmed parents had seen them stop breathing in their sleep for up to one minute and had been unable to rouse them. Under close scrutiny in the hospital, the doctors recorded similar events, the baby becoming completely unresponsive, the colour draining from its face and its body going limp. All three babies died within six hours of admission.

In the 1980s and 1990s, Belgian researchers studied the sleep patterns of more than 20,000 newborns. Those babies who became victims of SIDS experienced many more and longer sleep apneas than those who didn't. In many cases, babies where the cause of death is SIDS appear to have unusually muscular arteries for their age, exactly the sort of structural change to the cardiovascular system caused by frequent apneas in adults. The precise cause of death in SIDS is different from case to case, but it remains likely that sleep-disordered breathing of one kind or another explains some of these tragedies.

There is pretty good evidence too that bad breathing during sleep could play an important role in the onset of type 2 diabetes. Each apneic event is associated with an injection of adrenaline, an alarm hormone associated with preparing the body to fight or take flight – precisely what's needed if you've stopped breathing. Adrenaline has many functions, such as increasing the heart rate, dilating the pupils and opening the blood vessels to the muscles. It also acts on the pancreas, releasing hormones that cause a sudden injection of glucose into the bloodstream. In a fight-or-flight scenario this is useful, as the sugar rush provides the energy for muscle contraction. But if it happens hundreds of times throughout the night, it is the physiological equivalent of crying wolf. The cells around the body become resistant to the pancreatic hormone insulin, reluctant to follow its orders to remove glucose from the blood. This is type 2 diabetes.

Sleep apnea could also increase the risk of cancer. In the Wisconsin Sleep Cohort Study, for instance, those with moderate apnea were twice as likely to die from cancer as were those breathing normally during their sleep. For the most severe apneics, the cancer risk was almost five times above normal. The thinking here is that the low levels of blood oxygen that result from apnea cause something of a metabolic panic at the cellular level, increasing the risk that a cell will turn cancerous.

As if the thought that sleep apnea could be behind many cases of cardiovascular disease, SIDS, type 2 diabetes and cancer weren't alarming enough, there is also the fact that the resulting sleepiness makes accidents more likely.

\* \* \*

There have been many studies that have looked at the role of sleep apnea in road traffic accidents, and they all indicate much the same thing: if someone with sleep apnea is behind the wheel, they will typically have around twice as many accidents as those without. In the year 2000 in the US, it has been estimated that more than 800,000 drivers were involved in apnea-related car crashes, resulting in the loss of 1,400 lives, not to mention the estimated cost of $16 billion to vehicle and property damage and lost productivity.

Sleep apnea is not just a danger to road users. In 2013, Metro-North engineer William Rockefeller fell asleep at the controls of a train in the Bronx and, travelling at 82 mph, entered a bend with a speed limit of 30 mph. The derailment killed four people and injured almost 60. After the accident, an overnight sleep study revealed severe obstructive sleep apnea, with over 52 events per hour.

Writing up the case, the chief medical officer of the National Transportation Safety Board Mary McKay noted that between 2000 and the Bronx accident in 2013, the NTSB suspected sleep apnea played a part in at least ten accidents involving modes of transport from an oil tanker to an aeroplane. 'Preventing the next sleep apnea-related accident requires ongoing attentiveness throughout the medical care system to ensure transportation operators are screened, evaluated, diagnosed, and effectively treated for this condition,' she wrote.

\* \* \*

It was only when I began to learn more about sleep apnea that I became aware of similar stop-start episodes in my own sleep. The realisation came in stages, and it began with my father. It was February 2016 and we were on holiday with my parents, staying in a sixteenth-century thatched cottage in Sussex. In the hours before dawn, finding myself unable to sleep and not wishing to disturb my wife any more than my twisting and turning had already done, I got out of bed and went downstairs to read. In the sitting room, the remnants of a fire were still smouldering in the hearth. I switched on a lamp, pulled a rug around my shoulders and began to read.

It was very quiet, except for gentle waves of snoring billowing down from my parents' room above. Then they stopped. I looked up from my book. All I could hear was the blood pulsing through my ears. After 20 seconds or so, I heard a snort and several short, sharp intakes of breath, each imbued with a sense of panic, but one that diminished as my father's breathing settled back into its habitual snore. A few minutes later, the same thing happened. And again, and again. I had no idea my father suffered from sleep apnea, and this was the first time I'd heard the characteristic pattern of apneic breath.

It hadn't occurred to me that I might have sleep apnea too. But the following night, with the necessary first-hand knowledge, I suddenly recognised the unmistakable apneic pattern to my own nocturnal breathing. One moment I'd been asleep, then I became aware that I was hungrily gasping for air. The first time this happened I didn't think too much of it, but after the second, then third episode, it dawned on my transitioning consciousness that I'd just stopped breathing. The panting was accompanied by a mild but sickening sense of dread, rather like the sensation of going too long underwater at the swimming pool. Would I have picked up on this had I not just been reading about Etienne or been alarmed by my father's breathing the night before? Probably not.

Christian Guilleminault does not need to hear this story to know that I am at risk of sleep apnea. Towards the end of the interview, I suddenly notice he is staring straight at me. 'You are a mouth breather,' he observes. I realise that I am breathing through my mouth rather than my nostrils and I snap my lips shut. It's too late. Guilleminault has already computed my body mass index and, together with my gender, my age, my neck diameter, the shape of my face and my propensity for mouth breathing, he has assessed the probability that I suffer from some kind of sleep-disordered breathing. This, he admits, is something he does with almost everyone he meets, almost without thinking.

'If you are a mouth breather, you are not breathing well,' he says. 'You have a chance to become a snorer.'

'I snore,' I admit, but Guilleminault already knows this.

'On top of that you have your flat face and your maxillary deficiency.'

Now he's just being rude.

'Breathing is fluid dynamics. It's an issue of resistance,' he says.

In fact, snoring, the Pickwickian Syndrome and sleep apnea are really just different takes on the fluid dynamics issue. There are others. For a start, there's the lot you are born with. In the natural scheme of genetic variation, some people will just have airways that are more prone to collapse, like those with especially hefty tonsils. Others have what's referred to as upper airway resistance syndrome, where the airways do not collapse but they are particularly narrow and the extra effort to breathe interferes with sleep. Having asthma doesn't help.

Fetal and neonatal development can also have a huge effect on the resistance of your airways. Guilleminault and colleagues are currently following some 400 babies born prematurely in Taiwan. By the age of five, almost 80 per cent of these children appear to have problems breathing during sleep, he says, and the more premature the baby, the more likely they are to suffer. 'During the last three months of gestation the fetus continuously trains reflexes – particularly sucking and swallowing – that lead to continuous training of the muscles of the mouth and face,' he says. Without this workout, a newborn baby may end up with weaker face muscles and a smaller mouth, nose and upper airway than they might have had if the pregnancy had gone to term. It's important that the workout continues during the first few years of life too.

Then there's the obesity issue, a huge and expanding problem across the developed world. In the US, for instance, one in three adults is obese. The UK is near the top of Europe's obesity league, with one in four adults boasting a body mass index of 30 or more. It is easy to see that this should increase the risk of sleep apnea. 'When you are fat, you put your fat in your tongue and you reduce the size of your airway,' says Guilleminault.

What is less obvious is that there can be a vicious cycle, with sleep-related breathing exacerbating obesity and so on. There are probably several reasons for this. For a start, people who struggle to breathe at night tend to be very sleepy during the daytime, which doesn't exactly burn up the calories. As nineteenth-century physician Robert Macnish put it so unkindly in *The Philosophy of Sleep*, those with 'a disposition to doze on every occasion' have 'passions as inert as a Dutch fog, and intellects as sluggish as the movement of the hippopotamus or the leviathan'. Then there's the evidence that sleep apnea is associated with major metabolic changes, messing with the expression of genes involved in circadian regulation, sleep homeostasis and metabolism, and with the normally precise regulation of glucose in the bloodstream in a manner that is characteristic of obesity. There may also be a role for ghrelin and leptin, a pair of hormones with a good-cop-bad-cop influence on hunger and satiety (the feeling of fullness). With sleep deprivation, ghrelin rises and leptin falls, hunger goes up and satiety goes down, with crazy consequences for food consumption.

So suffering from sleep apnea is a little like swimming round the edge of a whirlpool. With stop-start breathing, body mass rises, which exacerbates the apnea, which nudges the BMI still further towards obesity and so on. There's a real risk of being sucked into the vortex.

It's not just people with sleep apnea who face this danger. In fact, where there's a problem with sleep, obesity almost always becomes an issue. This is certainly the case with narcolepsy, where obesity is more than twice as common as it is in the general population.

When Masashi Yanagisawa and colleagues first knocked out the hypocretin system in mice, the narcoleptic rodents became alarmingly fat. In the first paper to make this observation, there is a photograph of two mice, a normal mouse and its super-sized, knock-out littermate. With a bulging frame resting on tiny feet, it looks like a character from a kids' cartoon. Paradoxically, the knock-out mice did not obviously eat more than control animals, yet still almost doubled in weight. When under observation, people with narcolepsy don't obviously eat more either.

I used to be slim. Then, in my early 20s, when I should have stopped growing, I began to expand. Three wonderful summers I spent in Hungary doing fieldwork on animal behaviour, being overfed with delicious foods like szalonna (a smoky, fatty bacon), hurka (paprika-spiced blood sausages) or pörkölt (a rich and meaty stew, my favourite) did not help. But it's very likely that narcolepsy, which I'd just developed, also played a part in my transformation from a body mass of 75 kg to one that I struggle to keep under 90 kg. This equates to a BMI hike from 23.1 (healthy) to 27.8 (overweight).

The same factors that push someone with sleep apnea towards obesity – the inactivity, the metabolic changes and hormonal disruptions – are almost certainly at play here. But it's more than likely that the absence of hypocretins is part of the problem too. It's clear, for instance, that ghrelin and leptin exert some of their influence through the hypocretin system, so in its absence, the tight grip these hormones usually have on food consumption is likely to be considerably looser.

Returning to the theme of this chapter, the rising body mass index of many with narcolepsy threatens to add some kind of sleep-disordered breathing to their list of woes. This can complicate diagnosis. The presence of sleep apnea can mean that narcolepsy gets overlooked. Conversely, those with a diagnosis of narcolepsy and an explanation for their relentless daytime sleepiness may not notice the strangling grip of apnea.

\* \* \*

Given the dangers of apnea, it would help to have a simple way to assess the level of risk. The 'STOP-Bang' questionnaire is useful in this regard: eight simple questions about snoring, tiredness, observed apneas, blood pressure, body mass index, age, neck circumference and gender. It was devised by anaesthetists to screen prospective patients for erratic breathing during sleep that might cause complications during and after surgery.

Do you snore loudly? Do you often feel tired or sleepy during the daytime? Has anyone observed you stop breathing during your sleep? Do you have or are you being treated for high blood pressure? Is your body mass index greater than 35 kg/m$^2$? Are you over 50? Is your neck circumference greater than 40 cm? Are you male? If the answer is yes to three or more of these questions, there is a 90 per cent chance that the patient has some degree of apnea, and the anaesthetist will insist on further investigations before the operation can take place.

I snore, I am often tired in the day, I have stopped breathing in my sleep, I have elevated blood pressure and I'm male, so according to the STOP-Bang rules, I have high risk of sleep apnea. Guilleminault's observation of my mouthbreathing, flat face and deficient jawline led him to much the same conclusion.

Based on the most recent overnight sleep study I performed I experienced apnea, but only twice in the whole night, so I do not even qualify as a mild apneic. I did snore, for 26 minutes, or 7 per cent, of the night. This isn't enough to warrant medical attention, but it should not be ignored. Where snoring is loud, the vibrational energy could damage the blood vessels to the brain, increasing the risk of stroke. There is also evidence it could impair the hearing of anyone who gets too close, particularly in the ear closest to the noise. But even if it's only occasional snoring and not very loud, it should be taken seriously. People who snore are at a greater risk of developing sleep apnea and apnea can kill.

There are lots of ways to tackle snoring and obstructive sleep apnea. The most obvious is to lose weight. Rather less obvious is to take up the didgeridoo. Alex Suarez is a martial arts teacher and physical trainer living in Switzerland, and one of his students brought his didgeridoo on a week-long course he was running on the Mediterranean island of Ibiza. At the time, Alex had just been diagnosed with obstructive sleep apnea, and his doctor had explained that the back of his throat was in poor shape. He watched his student's circular breathing and hit upon an idea.

Alex bought a didgeridoo and began to develop a special set of exercises hoping to reduce the fatty deposits at the back of his throat and firm up the connective tissue and muscles. Within weeks of performing this customised buccal workout he'd completely silenced his snoring and cured his apnea.

Suarez's sleep specialist at the University Hospital of Zurich was so impressed that together they recruited 30 apneic patients and subjected the 'didgetherapy' to a randomised clinical trial. Half of the volunteers received an acrylic didgeridoo, lessons in circular breathing and instructions to practice for at least 20 minutes a day over the course of four months. The other half was put on an indefinite waiting list and given instructions to avoid any contact with a didgeridoo until an instrument became available. In the musical group, the patients experienced a significant improvement in their apnea-hypopnea index and daytime sleepiness. As a bonus, their bed partners found themselves less disturbed during the night too. There is no mention about how they felt about having a didgeridoo in the house.

If playing the didgeridoo seems a bit too much, there are other, more conventional measures that a low-level snorer like me can take to stave off the descent towards apnea. Kátia Guimarães and colleagues in Sao Paolo in Brazil have taken a series of exercises developed for speech therapy and demonstrated that they can reduce neck circumference, snoring intensity and the severity of obstructive sleep apnea, with substantial improvements in the quality of sleep at night and wakefulness during the day. There is a video in the online supplement to Guimarães' paper, showing about a dozen exercises variously designed to tone the tongue, the soft palate and the facial muscles. They involve a toothbrush, a balloon, some guttural utterances, plenty of sucking and some side-to-side jaw movement. I tot up the time that is needed to complete the requisite exercises and it's a good half hour a day.

\* \* \*

If it's too late for interventions like these it's time to turn to the continuous positive airway pressure (CPAP) machine. It first appeared in the 1980s, the brainchild of Australian physiologist Colin Sullivan. In essence, it is just a fan, a tube and a face mask. The mask is strapped to the patient's face at night and the fan blows a continuous stream of air into the lungs. It's sufficient to keep the airways open but not so strong that the patient wakes or can't exhale.

Before experimenting with the device on humans, Sullivan had been studying dogs, specifically brachycephalic breeds like pugs and British bulldogs well known for their operose breathing. It was no mean feat to fit a facemask to a dog, but the canines passed a snortless night, had a much better balance of blood gases and were far more active the following day.

It was time to try out the CPAP tech on humans. 'The immediate clinical response to one night of unobstructed sleep was remarkable,' wrote Sullivan and colleagues of their initial trials. 'Each patient awoke spontaneously, was alert, and remained awake unprompted for the rest of the day.' One of Sullivan's first subjects suffered from such serious apnea that he couldn't stay awake for more than a few minutes at a time. 'After the test night he remained awake for the entire day and was able to watch television for several hours – something he had been unable to do for years.'

Sullivan had no idea how big his invention would become, but one of the five patients in the trial of the new device had a hand in taking CPAP to the next level. In February 1981, just before the publication of Sullivan's paper, Eddie M. came to him and explained that he wanted his own machine at home. 'At that time, there were no suitable masks; they were either too heavy, or more importantly could not maintain a seal,' Sullivan remembers. By attaching the motor from a vacuum cleaner to a custom-made face mask moulded from dental adhesive, Eddie M. became the first person to harness the powers of CPAP from the comfort of his own bed.

'He was failing at work; his apnea caused sleepiness and compromised his ability to function,' wrote Sullivan, years later, of this patient. 'On treatment he went on to manage a very large corporation.'

Bill Dement has nominated Colin Sullivan for a Nobel

Prize for Physiology or Medicine. Twice. The invention of CPAP, he wrote in 2009, 'is rivalled only by the development of penicillin'. This might sound like an exaggeration, except that this single invention will have saved many millions of lives, preventing avoidable accidents and reducing the incidence of untold numbers of life-threatening pulmonary and cardiovascular events.

Sleep apnea is significant for another completely different reason, argues Kenton Kroker, a historian at York University in Toronto, Canada, and the author of *The Sleep of Others*. 'Sleep apnea really created sleep disorders medicine as a stable field.' In the early 1980s, there were just 34 dedicated sleep labs in the US. By the end of the decade, with sleep apnea on everyone's radar and CPAP promising treatment, there were more than 1,000, he says. 'It made sleep a public object for the very first time.'

The first commercial CPAP device appeared in 1983, a 60W 240V AC motor, driving a vortex blower designed for a spa-bath. It weighed almost 7 kilos and was so noisy that it had to be located some distance from the bedroom. Things have obviously changed since then, with new inventions making CPAP machines smaller and quieter and a micro-CPAP – 'the world's first maskless, hoseless, cordless' CPAP device – in the pipeline, yet there are still plenty of people who feel it's a little like sleeping with a vacuum cleaner, and they don't stick with the treatment. This can be dangerous, because the apnea comes back immediately.

In years to come, it's likely there will be alternatives to CPAP. For instance, there's an electronic device in development that picks up apneic events and delivers a small shock to the hypoglossal nerve in the neck, gently tugging the tongue back into position and opening the airways. All this is achieved without waking the patient. In an early trial, this approach proved remarkably successful, dramatically reducing the number of apneas and daytime sleepiness.

Improvements in CPAP and new technologies like these are all well and good, but they won't do much if people are not made more aware of the evident dangers of sleep-disordered breathing. In July 1967, over a decade before Colin Sullivan developed the prototype of what would become the first CPAP machine, he woke up to silence. This was odd. As a medical student he was still living with his parents and was used to the sound of his mother's snoring. When he went into her room, he found her dead. 'I did not know then, but I know now that sleep apnea contributed to her death,' he wrote in the 1990s.

# 7

# The perfect neurological storm

*'The brain is the most fascinating object in the universe.'*
Stanley Pruisner

Let's play detectives.

The crime is the theft of my hypocretins. My goal is to find who (or what) dunnit. I print a photograph of the victims (the hypocretins) and pin it onto the centre of the cork board above my desk.

In humans, it's exceedingly rare to find mutations in the genes encoding one or other of the hypocretin receptors (Mignot's Dobermans) or the hypocretins themselves (Yanagisawa's knockouts). Sometimes, however, narcolepsy does run in families. The most striking case of 'familial narcolepsy' comes from Spain.

In 2005, Andrés Lopez-Bartolí's doctor referred him to the Sleep and Epilepsy Clinic at the Gregorio Marañón University Hospital in Madrid. It looked like he might have narcolepsy, but it fell to sleep specialist Rosa Peraita-Adrados to carry out the formal diagnosis. Based on his history of sleep episodes and cataplexy, Peraita could be 'pretty sure' that he had narcolepsy without doing the tests, but the sleep study clinched it.

Andrés' story might have ended there, except for a throw-away comment. 'He told me that he had a twin brother with the same disease,' recalls Peraita. This was interesting. Over the course of her career – some 30 years at that point – Peraita had seen hundreds of patients with narcolepsy. In a few families, more than one sibling might be affected. 'But it was the first time in my life that I had twins,' she says.

When, after much persuading, Julián agreed to come and see her, she was in for an even bigger surprise. 'He told me he had two sons with the same condition,' she says. 'I was absolutely excited.' With further probing, it became clear that there had to be a genetic explanation of narcolepsy in this family. Andrés and Julián's mother Rosa had clearly suffered, as had her father Domingo before her. 'I decided then to study the whole family,' says Peraita.

In Spain large families are common, and the Lopez-Bartolí clan is typical in this regard. When Rosa (the mother) died a decade earlier, she'd left a lot of children. In addition to Andrés and Julián, there were four other siblings – one boy and three girls. But when Peraita asked after them, she encountered an unexpected problem. Neither Andrés nor Julián could tell her where any of their brothers and sisters were. Nor could they help Peraita locate any of Rosa's own siblings, which might have opened up other branches of the family tree for investigation.

Francisco Franco has a lot to answer for. In the middle of the twentieth century, Spain's fascist dictator maintained control by rounding up dissenters, and Rosa and her husband – both outspoken communists – wound up in prison, leaving six young children to be raised by uncles and aunts. Andrés and Julián stayed in touch, but lost contact with their other siblings. Most of them, they assumed, had probably passed away.

Just when it looked like Peraita would have to settle for studying a family of four narcoleptics – Andrés, Julián and Julián's two sons David and Miguel-Angel – something incredible happened. Casting her gaze down the list of new patients at her clinic one day, Peraita's eyes came to rest on the name of a suspected case of narcolepsy: Jesús Bartolí-Maqueda. 'It was a miracle,' she says. 'I saw the name of the family and I realised they had to be related.' Peraita began to feel like she was the host of some populist TV show, reuniting long-lost members of an estranged family. In the sterilised corridors of the Gregorio Marañón University Hospital, everyone was talking about Peraita's 'family'.

With Jesús' help, Peraita began to fill in the other side of the family tree, and found narcolepsy everywhere. Jesús' mother Teodora had been known as *la dormida*, 'the sleepy one', and Rosa, Teodora and another sister, Patro, had been dubbed *las bellas durmientes*, 'the sleeping beauties'. One of Jesús' sisters, Josefina, received a positive diagnosis, and it looked like several of her children might be afflicted too. But here again, Peraita ran into difficulties. All four of Josefina's sons were in prison for drug trafficking and consumption.

'I called the physician in the jail and told him I was studying this family,' she says. Armed with the appropriate paperwork, Peraita entered the prison to collect tissue samples. Two of the four brothers, and two sisters (not in prison) turned out to have narcolepsy.

Jesús even had a hunch that Antonia, one of Andrés and Julián's long-lost sisters, might not be dead after all. 'She's probably living in Avila, 150 km from Madrid,' he told Peraita. By making some calls to the local authorities, she succeeded in tracking down Antonia and drove out to Avila to meet her. 'Andrés and Julián didn't know their own sister was alive,' she says. As it turned out, she didn't have narcolepsy, but Peraita had revealed another branch of the family tree.

After six years of sleuthing, Peraita had identified 13 members of the family living with narcolepsy and, in collaboration with geneticists in Switzerland, she identified the genetic mutation responsible. The single gene that separates the narcoleptics from the non-narcoleptics in this family encodes a key protein that is essential for the proper development of certain brain cells. Interestingly, the same protein has also been implicated in a range of other disorders, including depression, bipolar disorder, schizophrenia and multiple sclerosis, though the reason why this should affect the production of hypocretins and lead to narcolepsy remains unclear.

I print out the family tree that Peraita and her colleagues published in their paper and I annotate it with the real-life names of this extraordinary family. I stick it to the cork board.

It's extremely rare to find narcolepsy being caused by a straight-forward genetic mutation like this. There are occasional clusters of cases within families, suggesting that for some people there is a pretty strong genetic basis to the disorder. It's possible, in fact, that the nineteenth-century bookbinder Herr Ehlert had an inherited form of narcolepsy, because he mentioned to his doctor Karl Westphal how his 60-something mother 'at times falls asleep while performing ordinary chores'. But in general familial narcolepsy accounts for less than 5 per cent of cases.

* * *

Some cases of narcolepsy appear to be caused by physical damage to the brain. Jean-Baptiste Gélineau clearly quizzed Monsieur 'G' the barrel-maker about head injuries, because the patient described a heated argument during which he'd sustained a violent punch, and a second incident where 'a log fell on his head'.

Head injuries like this are extremely common. It's been estimated, for instance, that around 80,000 people every year in the US need medical care as a result of a wallop to the head from car crashes, falling over, violent assaults or sports injuries. A fair whack of these people end up with some kind of sleep disorder, be it post-traumatic hypersomnia, sleep apnea, narcolepsy or insomnia.

Dorothy Ennis-Hand's narcolepsy was probably caused by an accident, one that occurred on a bitterly cold night in Dublin in 1953 a few years before leaving Ireland for England. She was on the way to see the recently released western musical *Calamity Jane*, riding on the back of her brother's motorbike, when Dorothy was thrown from the vehicle, landing helmetless on her head. 'I remember for a split second actually seeing stars.'

When the ambulance arrived, Dorothy's face was covered in blood, and the paramedics were reluctant to remove the scarf she'd wrapped around her head for warmth. It turned out that the real damage was far more profound than the superficial cut to her forehead. Within a few months, Dorothy began to feel the sleepiness that would later become diagnosed as narcolepsy.

In 1941, Wilson Gill – a physician to the North Staffordshire Royal Infirmary – described three cases of excessive sleepiness that appeared to have followed immediately after a head trauma. One was a 14-year-old girl, who was struck by a piece of lead falling from a skylight above her. In another, a 16-year-old girl, an electrical tester, was struck by a 2 lb insulator. The third, a district nurse, walloped her head on an oak beam while attending a patient. Not long after these traumas, all three women became excessively sleepy and experienced weakening fits that sound a lot like cataplexy. 'It is fair to assume that the accident to the skull was the responsible factor,' wrote Gill.

Some people have had miracle recoveries from what were clear cases of narcolepsy. But they had tumours or cysts that appear to have been pressing in on the brainstem, and when treated – either by chemotherapy and radiotherapy or with the surgeon's knife – the symptoms of narcolepsy vanished within a matter of days or weeks.

\* \* \*

For the vast majority of people with narcolepsy, they will be the only ones in their family to be afflicted and there will have been no obvious accidents at which to point the finger of blame. But genes and accidents still have a role to play in the majority of cases of narcolepsy, commonly referred to as 'sporadic' narcolepsy because of their unpredictable and isolated onset.

The colony of narcoleptic Dobermans inspired Mignot and others to look for a genetic signature common to all narcoleptics, and from the 1980s onwards evidence began to emerge for a strong signal coming from chromosome six.

This is of special interest because chromosome six is rich in genes involved in the recognition, handling and ultimate destruction of pathogens. The proteins produced by these genes – the so-called human leucocyte antigens (HLAs) – play a role in many autoimmune disorders, conditions in which the immune system misfires and starts attacking healthy cells. In type 1 diabetes, for instance, HLAs are implicated in the targeting of insulin-producing cells in the pancreas. Within days of this self-destruction, the body has irreversibly lost its capacity to manufacture this precious hormone.

It turns out that 98 per cent of people with a diagnosis of narcolepsy and cataplexy have an HLA in common, so perhaps narcolepsy – like type 1 diabetes – could be an autoimmune disorder, the body's immune system somehow running amok. The HLA in question (called DQB1*06:02) can only be part of the story, though, because DQB1*06:02 is very common indeed; roughly one in every four people boasts a copy, yet only a minute, unlucky fraction of these carriers live lives destroyed by sleep. 'There is clearly a genetic predisposition to narcolepsy,' says Mignot. 'But it was also obvious from very early on that there had to be one or more environmental triggers.'

This makes sense because I, like the vast majority of people with narcolepsy, have not always been this way. In fact, narcolepsy is most likely to strike during adolescence.

The clearest evidence for this comes from a study of almost 350 people with narcolepsy, some from France, some from Canada, in which researchers plotted out the age at which the symptoms of narcolepsy appear. In both populations, there was a significant peak in the onset of narcolepsy at around 15 years old.

Why would this be? One idea is that narcolepsy is linked to some kind of stress and stresses – either physiological or psychological – which are a notorious feature of adolescence. Interestingly, there is also a smaller, secondary peak in the onset of narcolepsy, one that shows up around the age of 35. Stress could conceivably play a role here too, for mid-life is well known for its sense of crisis.

Given that the immune system appears to play some part, however, the most likely stressor is an infectious disease. Researchers began to point the finger of blame at two pathogens in particular: *Streptococcus*, the bacterium responsible for 'strep throat', and influenza, the viral agent of flu.

Interestingly, an infectious agent is implicated as the trigger for a mysterious outbreak of sleepiness in Europe towards the end of the First World War. Encephalitis lethargica, or von Economo's disease, is thought to have affected around five million people at the time, with symptoms of fever, sore throat, lethargy and in severe cases a coma-like state from which many never emerged. The precise pathogen remains unclear, but it's possible that the outbreak of Spanish influenza in 1918 had something to do with it.

Since infectious agents give us more of a runaround during the cold winter months, when immune systems are at their weakest, we might expect to see some sort of seasonal pattern to the onset of narcolepsy.

Beijing University People's Hospital is a big hospital, and its doctors typically diagnose dozens of new cases of narcolepsy every year. When they asked people with narcolepsy to identify the month in which they'd first noticed their sleepy symptoms and plotted out their findings, they discovered a striking pattern. The graph rises and falls like a childish depiction of the sea, up and down, up and down over a 15-year period from 1996 to 2011. Every April there's a peak in patients noticing symptoms of narcolepsy and every November there's a trough.

I think back to the onset of my own symptoms, more than 20 years ago. It was April 1994, or thereabouts. Did I suffer a nasty infection that winter? In all honesty I don't remember. But I do know that my girlfriend at the time battled through November 1993 with a horrific strep throat, so I would certainly have encountered these same pathogens too.

I add a double helix and an electron micrograph of *Streptococcus* to 'the board' to indicate the genetic predisposition and a subsequent infection. But taking a step back, I'm a little uneasy with the evidence gathered so far. All of it is circumstantial, nothing establishing a causal link between an infection and narcolepsy. What would really advance the case for the prosecution?

An experiment would be nice, with a large number of volunteers assigned at random to one of two treatments. One group would be injected with a suitable pathogen, influenza virus perhaps, and the other with a control, like a 0.9 per cent saline solution. If influenza really does somehow trigger the onset of narcolepsy, we'd expect to see more cases of the sleep disorder emerging in the flu group. It's that kind of evidence that might swing a jury.

Such an experiment, of course, could never be done. Who is going to sanction a trial where the aim is to establish a link between an infectious disease and a disabling neurological disorder? Answer: no one. It turns out, however, that a wild strain of influenza that appeared in 2009 set the stage for a natural experiment with an unfeasibly large sample size along exactly these lines.

On 12 April 2009, the authorities in Mexico reported an outbreak of an influenza-like illness in the small rural town of La Gloria in the state of Veracruz. A few days later, with more cases emerging in other states, officials were on high alert, carefully monitoring the spread of the disease. Genetic analyses soon identified the pathogen: a strain of influenza normally present in pigs appeared to have jumped to humans. It became known as swine flu.

This sort of thing happens all the time, but usually only causes isolated cases in humans. One person gets ill and recovers. The 2009 outbreak was different. The virus was being coughed from one person to the next. By 23 April, officials had ordered the closure of schools in Mexico City and soccer fans were being refused entry to games, the matches taking place but in empty stadia. By the end of April, there had been reports of serious flu-like disease in every one of Mexico's 31 states, with 1918 suspected cases and 84 deaths.

By then, geneticists had identified the strain of the virus as H1N1, the very same as the Spanish influenza responsible for the global flu pandemic in 1918, estimated to have killed somewhere between 50 to 100 million people. With cases of swine flu being confirmed across the world, the World Health Organization declared a global pandemic in June.

In short, in the middle of 2009 there was intense institutional fear that the swine flu could bring about a repeat of the Spanish flu. Pharmaceutical companies were soon bidding for multimillion-dollar contracts to rush out a vaccine before the flu season really got under way in the winter months of the northern hemisphere.

From September onwards, several different vaccines – all designed to protect against swine flu but subtly different in their manufacture – began to roll off production lines around the world. One of them, it turned out, would result in a small but significant increase in the incidence of narcolepsy. The others would not. Awful as this is, it's as close to a controlled experimental test of a link between infectious disease and narcolepsy we could ever hope to see.

Markku Partinen remembers the moment he got the first inkling of an association. He is a consultant neurologist in Finland and director of research at the privately run Helsinki Sleep Clinic, a handsome man with a generous sweep of once-blonde, now-silvered hair. In February 2010, he was seeing patients at his sleep clinic. There were several new patients on his list that day. One of them, a seven-year-old boy, just could not stay awake for more than an hour at a time, even after what appeared to be a good night's sleep. 'It was very severe and had started very abruptly in December 2009,' Partinen tells me. 'It was quite clearly a case of narcolepsy, but the sudden onset was unusual.'

Partinen immediately thought of H1N1. 'There was an epidemic going on so everyone was thinking about the H1N1 influenza.' He asked if the boy had suffered a recent bout of flu. His parents couldn't be sure, but he had received the swine flu vaccine, they told him.

This could just be coincidence, but Partinen made a mental note. 'Then, during the spring, not a long time after I saw this patient, there was another child, again with abrupt narcolepsy,' he recalls. 'Then there were more and more.' By June, he had seen around five times as many new cases of narcolepsy as he'd normally see in a year, and most of them were children. He made a point of asking about swine flu. 'The common denominator in all cases was the vaccination,' he says. 'That was really the key.'

\* \* \*

At about the same time that Partinen saw his first case in early 2010, Josh Hadfield began to show signs of excessive sleepiness. He was four at the time.

'It was very sudden,' his mother Caroline tells me. 'Very sudden.' It is 2015. Josh is ten, about the same age as my eldest son, and I have travelled to Somerset to meet him. As we get chatting about what it's like being at school with narcolepsy, I recognise the kind of beyond-his-years maturity of a boy who has been through a lot.

The sleepiness began soon after Josh had started in reception at primary school. Perhaps the change in routine was taking its toll, thought Caroline. But when she reflected, she knew it had to be something else. During his nursery years, Josh had been able to go without sleep for 12 hours a day without difficulty. This was completely different. 'He was coming home from school at 3.15, he was in bed by 4 p.m. and he would then sleep the whole way through until morning.'

Caroline and her husband got Josh through the last week of the February half term, expecting him to pick up during the holiday. If anything, however, his need for sleep became even more pronounced. He'd get up in the morning, be awake for about 45 minutes, then go back to sleep for another two hours, a pattern that continued throughout the day. 'He spent more time asleep than he did awake,' says Caroline. Sleeping through the night and most of the day, Josh was probably awake for less than six hours in 24, she says.

When Caroline sought a medical opinion, her GP ran some tests for glandular fever, a viral infection that is a suspected risk factor for myalgic encephalomyelitis or chronic fatigue syndrome (ME/CFS). This drew a blank, but it wasn't long before there was another alarming development. Josh was at home watching a *Tom and Jerry* cartoon, when his head slumped against the sofa and his eyes rolled back in their sockets. 'It looked like he was having a fit,' recalls Caroline, unaware at that time of the existence of cataplexy. The ambulance arrived within five minutes, but by then Josh had recovered, fallen asleep and woken up. Still, on the paramedics' advice, Caroline took him to the local hospital.

This didn't help. 'There's nothing wrong with him,' the doctor told her. 'He's just not sleeping properly. You need to get into a sleep routine.'

In fact, the trigger for Josh's sleepiness and strange fits was a vaccination. Caroline had received a letter from her GP in late

2009 with an appointment to have him immunised against the H1N1 swine flu. 'Because he was under five, he was classed as being at risk,' she says. She did her research 'as any parent would do', but the only side effect appeared to be a bit of a sore arm and occasional flu-like symptoms. This was 21 January 2010.

Caroline knew of several families in her neighbourhood that had been struck by what appeared to be swine flu. 'It had been quite horrendous,' she says, and with this and the intense media coverage in mind, she was persuaded. 'Of course you think, "I've got to protect my son,"' she says.

Josh took the injection well. He suffered no soreness. No fever. Then, a couple of weeks later, he descended into a state of near-permanent somnolence.

* * *

Partinen, meanwhile, had initiated a collaboration with his counterparts in Sweden to test his hunch that the swine flu vaccine could be triggering narcolepsy. The incidence of the disorder had skyrocketed, he found, the increase particularly noticeable in very young children. Partinen and his colleagues estimated that prior to 2010, just one in around 325,000 Finnish children under the age of 17 would be diagnosed with narcolepsy every year. In 2010, the odds collapsed to one in 20,000. Of 54 patients in this age bracket diagnosed with narcolepsy in 2010, 50 had received Pandemrix, a vaccination against swine flu manufactured by British pharmaceutical giant GlaxoSmithKline.

In August 2010, a week after the World Health Organization announced that the pandemic was over, the Swedish authorities revealed they were investigating six cases of narcolepsy in children suspected to have been triggered by the swine flu vaccination. The Finns soon went public with their own evidence. In the case of Finland, Pandemrix was associated with a 13-fold increase in the risk of developing narcolepsy in children. 'We normally diagnose five to eight new patients a year,' says Partinen, talking of his clinic. Scaling this up for 2010 and it's close to 100 new cases, he says. 'That is what we found in the first year.'

As other countries began to investigate, a similar pattern began to emerge. During the pandemic, there were several different H1N1 vaccines in circulation. Only Pandemrix appeared to be associated with narcolepsy. In the UK, the government placed a massive order with GSK for millions of doses of Pandemrix. Josh Hadfield received one of these doses towards the end of January 2010.

\* \* \*

In the absence of a diagnosis for Josh, his mother Caroline and her husband cast around for a possible cause. 'We came up with everything,' she says. Top of the list was a brain tumour, but a CAT scan ruled that out. Josh had experienced most of his fits whilst watching TV, so perhaps the furniture – now about ten years old – had something to do with it. 'We went out and bought new sofas.' They considered lead poisoning. 'He had a lot of Thomas the Tank Engine toys and some of the old metal ones,' she says.

On one of many visits to the hospital, they happened to get a young doctor they hadn't seen before. Caroline told him how Josh had been having difficulty walking, collapsing a little to one side as he went. Josh made his way, lopsidedly, to the end of the ward and back. The doctor watched with interest. 'He then mentioned cataplexy,' remembers Caroline.

By August 2010, Josh had been referred to Zenobia Zaiwalla, director of the Oxford Sleep Centre at the John Radcliffe Hospital in Oxford. 'She took one look at him, turned around to me and said he's got narcolepsy and cataplexy,' says Caroline. She couldn't trust a diagnosis made on such scant evidence and questioned Zaiwalla. 'We will need to do a sleep study to confirm this,' she said. 'But that is what this is.' Then Zaiwalla, who was aware of the link that Sweden and Finland had made between Pandemrix and narcolepsy, asked if Josh had been vaccinated against swine flu. Caroline answered in the affirmative.

In 1979, the British government passed the Vaccine Damage Payment Act to compensate those suffering permanent disability as a result of a public vaccination campaign. 'If an individual undertakes an act on behalf of society, which is what a vaccination is, and then if they are harmed as a result of that, we should compensate them and look after them. That just seems reasonable and human to me,' says Matt O'Neill, chair of Narcolepsy UK. But in order to qualify for the one-off, lump-sum payment of £120,000, Josh and other British children who developed narcolepsy following the Pandemrix vaccination would have to demonstrate at least 60 per cent disability.

Owing to a quirk of history, this means demonstrating a disability equivalent to the 'amputation of leg below middle thigh, through knee, or below knee with stump not exceeding 4 inches'. As the injury that causes narcolepsy is buried deep inside the brain and the symptoms aren't always on show, this has not been straight-forward, and there has been a succession of judgments and appeals. Only a few of these children have now received the £120,000 compensation. In Josh's tribunal, the judge reviewing the evidence decided that, based on his present symptoms, he had been 72 per cent disabled.

The compensation will help, says Caroline, but it's not a lot for a life consigned to sleep. 'No amount of money in this world can ever compensate for what Josh is going through,' she says. 'The only thing they could do to make things better would be an out-and-out cure.'

Peter Todd is a solicitor for the London law firm Hodge, Jones and Allen, and is representing many British children who developed narcolepsy as a result of the vaccination (though not Josh) in a class action against GSK under the Consumer Protection Act. 'It's a question of social justice,' he says. There is also a danger that failing to acknowledge a genuine adverse reaction like this and pay fair compensation 'can undermine uptake of vaccines generally and lead to outbreaks of deadly diseases which otherwise could be successfully eradicated'.

\* \* \*

What was in Pandemrix that could have caused narcolepsy? There were three main ingredients: first, and most importantly, inert components of the swine flu virus, fragments of viral protein intended to prime the body's immune system should it encounter the real thing; second, there was an adjuvant, a bonus compound inserted into a vaccine to guarantee a strong immune response; third, it contained a preservative called thiomersal. In addition, Pandemrix, as all vaccines, will have contained a few trace elements, an inevitable consequence of the mass manufacturing process.

One of these ingredients could help us understand the cause of narcolepsy, not just in those affected by Pandemrix but also for narcoleptics like me. Partinen and his colleagues have studied samples of blood from those who received Pandemrix and developed narcolepsy. If, as is hypothesised, the GSK vaccine provoked children's immune systems in a manner that other vaccines did not, it should be possible to detect a unique signature in the suite of white blood cells and antibodies. Their findings suggest that the key difference between Pandemrix and other vaccines was not the adjuvant or the preservative but a very particular fragment of the H1N1 virus itself.

There are few upsides to the Pandemrix story, except that it gives researchers like Partinen some important clues as they seek to unravel the precise sequence of events that results in the autoimmune attack on hypocretin cells in the brain. If we knew this, it might be possible to spot an ongoing attack and do something about it before it's too late.

\* \* \*

I review the board. Andrés and the Spanish family show that in rare cases there are simple genetic mutations that can cause the disorder. Dorothy Ennis-Hand's motorcycle accident suggests that a head injury can do it too. But for most people with narcolepsy, it looks like a whole series of factors have conspired to trigger an autoimmune attack on any cells that just happen to be in the business of manufacturing hypocretins.

There are genes, DQB1*06:02 for sure, but almost certainly others we don't yet know about. Emotional stress may come into it. Maybe hormones in some way. An infection looks like it's probably critical. 'It's like the Swiss cheese model of accident causation,' says Mignot, referring to the extremely low probability that holes in randomly chosen slices of Emmenthal will somehow miraculously align. I think of it as the perfect neurological storm, contingent on many factors, highly improbable and completely devastating.

Adding further complexity to this increasingly elaborate

picture, Mignot tells me about one more element that may be required for the onset of this kind of 'sporadic' narcolepsy: chance.

Owing to the way the immune system is built, the cells in circulation at any one moment are the product of a unique sequence of unpredictable events. During fetal development and in the months after birth, there is random generation of a vastly variable population of white blood cells. Before being released into the body, these newly created immune cells pass through the thymus, an organ whose task it is to seek out and destroy any cells that might cause a problem by cross-reacting with the body's own cells. It's a little like a military recruitment in which there is a security check, with ruthless officers terminating any soldiers who could be working for the enemy.

When a cell makes it through the thymus, it gets deployed into the circulation. If it ever meets a foreign cell that it binds to, the immune system will generate a cohort of identical cells to bolster the defence. In the military analogy, this would be like cloning those soldiers with a proven track record, creating entire regiments of identical individuals. If an immune cell does not meet a pathogen, it either lies dormant or disappears altogether.

In narcolepsy, there is probably a small number of unlucky people who have immune cells that were not weeded out by the thymus and go on to cross-react with hypocretin cells. In effect, white blood cells get through the security check and end up taking out the hypocretin neurons in the brain.

'Why didn't the thymus do its job?' I ask.

'Bad luck,' says Mignot. 'A lot of people think that a condition like narcolepsy results from a simple combination of genetic and environmental factors. But in the case of the immune system, it's super-complicated.' Take identical twins. Even with the exact same genes, the exact same family history and the exact same infection, it's likely that their immune systems will have grown in different directions so will have different strengths and vulnerabilities. 'If you have a certain infection, you may fight it in a certain way just because a certain cell happens to be in the right place at the right time,' he says. 'After just a few infections, two identical twins will have different immune systems that continue to diverge throughout life ... The immune defence that you have right now is a product of every infection you have had and not had from before birth to the present.'

Intriguingly, there is a birth effect in narcolepsy, which means that people born in spring run a small but significantly greater risk of developing narcolepsy later in life than those born at other times of year. The same phenomenon is observed in other autoimmune disorders, though people with schizo-phrenia, Parkinson's disease, diabetes and Crohn's disease are all more likely to have been born during the winter months.

Why would this happen? The most likely explanation is that before or shortly after birth some seasonally variable factor conspires to compromise the immune system, unwittingly increasing the chances of developing a particular condition later in life. Anecdotally, my own experience tallies with this pattern. I was born in March, the most common birth month amongst people with narcolepsy. I ask my mother if she remembers being laid out with a bug in the winter of 1972 as I, inside her uterus, transitioned from my second to my third trimester. 'I had an awful cough that lasted a very long time,' she tells me. Her obstetrician told her it would clear up when I was born and it did. But not, perhaps, before predisposing me to a narcoleptic future.

I'll never know and it hardly matters now, but when Mignot refers to the role that bad luck probably plays in the development of narcolepsy, this is precisely the sort of serendipity he's talking about.

* * *

For those who have travelled the autoimmune route to narcolepsy (as I probably have), the prediction is that they will have lost most, if not all, of their hypocretins.

Thanks to two people with narcolepsy – a 77-year-old woman and a 67-year-old man – who donated their brains to science, it's possible to see what this looks like. In a section of brain from someone without narcolepsy that is stained to reveal hypocretins, there are somewhere in the region of 30,000 hypocretin-producing cells, sprayed like lead shot across twin spots in the hypothalamus, each about the size of a garden pea. In a person with narcolepsy, the stain doesn't stick at all.

Another way to check for the presence of hypocretins (and a method that can be performed pre- rather than post-mortem) involves tapping into the spinal column to draw off a sample of cerebrospinal fluid. In most people, this fluid is flush with hypocretins. In many people with narcolepsy, the hypocretins are absent. Interestingly, however, some people with a diagnosis of narcolepsy still produce some (though usually reduced) amounts of hypocretins.

The natural variation in the levels of hypocretins probably helps explain why narcolepsy – in spite of its singular pathology – is a spectrum disorder. But there are other factors too that account for the incredible variation in symptoms from one person to the next and the wildly variable response to medication. The fact that everyone's brain is structured as a consequence of a unique blend of genes, environment and chance guarantees that the same amount of hypocretin deployed in two different brains will play out differently in the many different neural networks it impinges on.

The flip-side of this is that a complete absence of hypocretin, as is the case for many people with narcolepsy, will have similarly variable consequences. When the hypocretin system goes down, it's possible that some of the downstream neural circuits will make an effort to compensate. Compared to controls, for instance, people with narcolepsy have almost double the number of neurons in a region of the hypothalamus that synthesises the neurotransmitter histamine.

I was diagnosed with narcolepsy before the discovery of the hypocretins, so I have never had my levels measured. As a high-functioning narcoleptic at the fortunate end of the spectrum, I wouldn't be surprised if I've still got a bit of the stuff knocking around my brain and probably more than most. But looking at the cork board, I don't feel particularly fortunate. I add that last all-important contributory factor: bad luck.

# 8

# Lost in transition

*'For the night is dark and full of terrors.'*
George R. R. Martin

I remember a Saturday job I took shortly after I'd developed narcolepsy but before I had a clue what was going on.

Waterstones in the City of London had closed its doors to the public to do a stock take. There were probably 20 of us in the store that day, each given a handheld scanner and tasked with logging barcodes on shelved books into the store's database. It was monotonous work and I kept finding myself zoning out, then suddenly startling back to consciousness.

I had absolutely no idea at what point I'd lost awareness of my surroundings or how long I'd been unconscious but nobody seemed to have noticed. My scanner was still pointing at the spine of a book so I resolved to carry on in spite of the worrying thought that I might be bleeping the same book twice or even three times. Within an hour, the Waterstones staff called everyone to the till. Errors seemed to be creeping into the database and they'd decided to call it a day, paying everyone off in full for the anticipated four hours of work. Mercifully, they did not identify the culprit, but I am pretty certain who it was. This was my first encounter with automatic behaviour, the talent of losing consciousness but still managing to perform some (usually repetitive) task.

For most people, the brain is either awake or asleep, and the transition between these states is so fleeting it doesn't really register. In a way, this is remarkable. We – and pretty much every other animal – have evolved with brains in which consciousness is controlled by the flip of a switch. As we saw in chapter 5, it's likely that the hypocretins play an important role in the stabilisation of this switch, keeping it firmly 'on' or 'off' rather than fizzing between states, blurring the normally clear distinction between wake and sleep.

Sleep disorders that occupy this interstitial space are usually referred to as parasomnias. There are some, like automatic behaviour, where the brain starts in a position of wakefulness and borrows some of the characteristics of sleep. In others, the brain can be fully asleep yet acquire aspects of wakefulness. Either way, the results underscore why evolution should have done its best to minimise the confusion, disturbance and danger of a parasomnic existence.

Automatic behaviour like I experienced in Waterstones is common in narcolepsy, but can also occur in epilepsy, schizophrenia, in response to drugs or medication or simply following sleep deprivation. It's usually the case that the brain slips into a light 'microsleep', most likely stage 1 non-REM, where the eyelids close, the eyeballs start to roll and a thin veil descends over consciousness. As the direction of travel is from wakefulness towards sleep and it involves losing your mind (even if only temporarily), it can be an unsettling experience. In narcolepsy, automatic behaviour is often an everyday occurrence.

\* \* \*

Dorothy Ennis-Hand identifies automatic behaviour as her

earliest symptom of narcolepsy. Once she'd recovered from the superficial injuries from her motorbike accident, she took up a job stringing rosary beads into necklaces at the Mitchell Rosary Factory on Marlborough Street in Dublin. 'The smell was terrible,' says Dorothy, because making the beads involved boiling down cow horns to soften them, cutting them into small beads and singeing a hole through the middle of each one. Armed with pliers and a roll of fine wire, Dorothy was one of 'the decade girls', workers tasked with stringing together beads in groups of ten. On the next workbench, other women assembled these 'decades' into necklaces before the final step of adding a crucifix. The work was so monotonous that from time to time Dorothy would look down to find she'd put together a big pile of many more than ten beads but with absolutely no memory of having done so.

The things that people with narcolepsy can do in this state are really quite remarkable. I remember during my descent into narcolepsy recovering consciousness in a lecture theatre at university to find that I had continued writing notes. The content – on the impressive water-conserving adaptations of the desert-dwelling jerboa – were still discernible, though not particularly neat. Since then, I have learned to avoid writing emails late into the night, for it's all-too easy to switch to auto-pilot and press 'send' on a poorly punctuated work of witless nonsense that can incorporate elements of dreams.

Massimo Zenti's automatic behaviour, as a young boy attending a Catholic School in Verona, Italy in the 1980s, caused something of a stir. In the mornings, Massimo found he was able to stay focused. In the afternoon he'd begin to zone out, but it's clear he wasn't completely asleep. His eyelids stayed open, his eyeballs rolled upwards, but he carried on writing. The priest who ran the school telephoned Massimo's mother. 'They thought I was possessed by the devil and recommended an exorcist,' he says.

French cyclist Franck Bouyer began to fall asleep during training, especially on long, monotonous stretches, not quite asleep but not quite awake either. 'It was like I was sleep walking. I'd wake up and not know where I was. I had taken the wrong route without noticing,' he says. Franck was fortunate to get a rapid diagnosis of narcolepsy, within about a year of onset of symptoms. But this presented a problem. Fearing that he posed a danger to himself and to other cyclists, the French Cycling Federation decided he could not race unless he was taking medication to control his symptoms. 'They were frightened that I'd fall asleep on my bike.' By taking a stimulant to combat the sleepiness, however, Franck found himself in breach of the World Anti-Doping Agency's strict rules on banned substances. Catch 22. 'Narcolepsy destroyed my career,' he says.

In a more domestic setting, automatic behaviour can cause considerable confusion, with objects vanishing and turning up in unexpected places. Sarah Jackson has opened letters, only to keep the envelope and dispose of the contents. 'A couple of days ago we were hunting for the butter,' she says. 'It turned up in the freezer.'

The combination of this zombie-like state and the fridge carries another serious risk: binge eating. In my own life, I'm ashamed to admit I can decimate a jar of nuts, vanish a cake or expunge a chocolate mousse before I've even noticed. In a recent study, Dutch researchers quizzed people with narcolepsy about their diet and some 55 per cent reported binge eating with a sense of lack of control. For one in four, the bingeing was so extreme that they met the criteria for a bona fide eating disorder.

In a follow-up experiment (involving crisps, salty biscuits, cocktail nuts, wine gums, skittles and M&Ms), people with narcolepsy showed considerably less discretion over what they popped into their mouths than did control subjects. In fact, they ended up consuming almost four times as many calories over the course of the experiment as non-narcoleptics. It seems likely that 'aberrant food choices' like these probably contribute to the increased prevalence of obesity amongst people with narcolepsy.

When automatic behaviour takes place in public, it can be embarrassing. Angela Wells recalls a time when she'd been to the supermarket for a couple of miscellaneous items and come to at the till with a full trolley. At first she figured she'd accidentally adopted someone else's trolley but it turned out she'd been rolling it up and down the aisles filling it with some seriously odd and exotic foodstuffs. 'Some I would not have known what to do with,' she says.

This kind of unconscious behaviour has even resulted in illegal activity. In the 1970s, Wendy Kaufman's automatic behaviour got her arrested. Prior to her 'offence', the 55-year-old factory worker and mother-of-two had suffered from excessive daytime sleepiness for some 25 years and had received a diagnosis of narcolepsy. On a typical day at work – on an assembly line lubricating machine parts with graphite – she experienced several 'episodes of reduced awareness'. On at least one occasion, the eagle-eyed inspector roaming the shop floor had rejected a series of over-lubricated parts and she often caught herself rubbing graphite on the same part over and over again.

Wendy would have this same experience whilst shopping. Once, she'd unwittingly filled her trolley to the brim with jars of pickles and wheeled it out of the store before coming to her senses. She'd walked back into the supermarket and unloaded the pickles without anyone noticing. Another time, however, she wasn't quite so fortunate, hauled up for attempting to exit a hardware store with a bunch of DIY tools without paying. As it was her first offence, she'd been put on probation. But her second offence – 'stuffing her purse with packages of meat of a sort that she would not ordinarily choose' – resulted in Wendy standing trial for shoplifting.

'In all the cases of shoplifting, the patient claimed to be unaware of her actions,' wrote the sleep scientists who studied her. 'She did little to hide her actions and took items for which she had little use.' The doctors performed several sleep studies, confirmed the narcolepsy diagnosis and judged that 'this patient's automatic behavior was part of the narcolepsy syndrome'. Armed with this expert opinion, Wendy's defence counsel was able to argue successfully against the charge of shoplifting. In spite of medication that vastly improved her symptoms, she remained fearful there might be a repeat of her retail-based antics and took the sensible precaution of never shopping alone again. 'The family has not noticed any further shoplifting by the patient,' her doctors wrote in the *Journal of Clinical Psychiatry* in 1979.

Although there is no loss of consciousness in cataplexy, it does involve an intrusion of a sleep-related state onto wakefulness, namely the loss of muscle tone normally reserved for REM. These episodes can be dangerous, particularly if there's a chance they can happen at the wheel (which effectively rules out driving), if there's a long distance to fall (as in marmot-watching from a ski-lift) or in proximity of water (in the bath, larking around in a swimming pool or fishing, to cherry-pick just three perilous $H_2O$-based scenarios). Collapsing when in custody of children, whether they are in utero or mucking around at the edge of a busy road, is alarming to say the least.

* * *

When wakefulness intrudes on sleep, the upshot is often similarly troublesome. As many of these parasomnias result in partial arousal during deep sleep, they are particularly common in young children (who experience relatively large amounts of stage 3 non-REM) and are often confined to the first few hours of the night (when this phase is at its most abundant).

Somniloquy, or sleep talking, is a case in point. In a study of several thousand Brazilian schoolchildren aged between three and ten years old, around half were reported to sleep talk at least once a year and around one in ten claimed to sleep talk every night. Most somniloquy doesn't make much sense and tends to clear up by the time that puberty hits. Occasionally it persists into adulthood and can be strikingly coherent too, as in the case of American songwriter and lifelong somniloquist Dion McGregor.

During the 1960s, McGregor's roommates made recordings of his nocturnal ramblings, many of which were subsequently released in a series of albums and transcribed into book form. These are so coherently incoherent that they have the ring of dreams. Indeed, the brainwaves of sleep talkers like McGregor indicate that they are probably existing in a hybrid state combining the vivid dreams characteristic of REM and the movement that is possible during non-REM.

Enuresis, or bed-wetting, is another parasomnia that's pretty common amongst children. George Orwell is perhaps the most eloquent bed-wetter there's ever been, and his reflections capture the distress that this can cause. At the age of just eight, the young Orwell was sent to boarding school and within a couple of weeks was wetting his bed. 'It was looked on as a disgusting crime which the child committed on purpose and for which the proper cure was a beating,' he wrote in *Such, Such Were the Joys*. Every night, he'd pray: 'Please God, do not let me wet my bed!' to no avail.

When, after several warnings, the young Orwell awoke again between clammy sheets he was sent to the headmaster, 'a round-shouldered, oafish-looking man, not large but shambling in gait, with a chubby face which was like that of an overgrown baby, and which was capable of good humour'. Not on this occasion. The headmaster flogged him, the words 'you dir-ty little boy' keeping time with the blows, until the riding crop snapped and the bone handle went flying across the room. 'Look what you've made me do!' he said furiously. Orwell's bed-wetting and the reaction to it left him profoundly confused, suddenly aware that it was possible 'to commit a sin without knowing that you committed it, without wanting to commit it, and without being able to avoid it'. The beating was a turning point. 'Life was more terrible, and I was more wicked, than I had imagined.'

*Pavor nocturnus*, also known as night or sleep terrors, are a frequent formative experience, where the child sits up in bed and screams. There are all the signs of an intense, adrenaline-fuelled fight-or-flight response, with thundering heart rate, sweating, wide eyes and pupils dilated, and there is little a parent can do to console. After the attack, which might last a matter of minutes, the child settles back to sleep, and in the morning they will have little or no recollection of the event.

Sometimes the episode will extend to somnambulism, or sleep walking. In most cases, these nocturnal wanderings are pretty uneventful. The ancient Greek physician Galen, for instance, noted that 'he once spent a whole night walking about in his sleep, waking only when he struck a stone in his path'. But stranger things can happen.

The Archbishop of Bordeaux reported observing a young scholar 'in the habit of getting up during the night, in a state of somnambulism, of going to his room, taking pen, ink, and paper, and composing sermons'. Once the student had finished a page, he'd give it a read-through and even correct mistakes. The Archbishop was suspicious and held a board up between the young man and his parchment, 'but he continued to write on, without appearing to be incommoded in the slightest degree'.

Perhaps the most famous sleepwalker is Lady Macbeth. She might be a work of fiction, but Lady Macbeth's guilt-wracked rovings around Dunsinane Castle are by no means implausible. In Act V, Scene I, Lady Macbeth's maid describes her mistress's recent behaviour:

> Since his majesty went into the field, I have seen her
> rise from her bed, throw her nightgown upon her, unlock her
> closet, take forth paper, fold it, write upon it, read it,
> afterwards seal it, and again return to bed; yet all this
> while in a most fast sleep.

It is at this point that Lady Macbeth enters, her eyes open 'but their sense is shut'.

If a sleep walker ends up in the kitchen, what began as an innocent perambulation can turn into something rather less savoury. In contrast to the binge-eating that occurs when sleep intrudes on wakefulness, sleep-related eating disorder tends to occur when in the deepest stages of non-REM sleep. There is little ceremony, the episodes of eating usually carried out at an open fridge or cupboard and at quite some pace. Cutlery is often superfluous, one woman happily using her hands to channel spaghetti and meatballs from a bowl to her mouth.

'The mess next morning is unbelievable, the empty boxes and food wrappers, the empty fridge ... a disgrace. I feel ashamed and I hate myself,' reported one woman to the researchers who first described the phenomenon back in the 1990s.

Sometimes people with sleep-related eating disorder will consume wholly inappropriate foods, like raw fish or bacon, butter and oil. But of all the cases I pore over, there is one woman who stands out, reporting an extraordinarily varied night-time diet of cat food, salt sandwiches, buttered cigarettes and odd concoctions prepared in a blender. Sometimes she turned on the stove and then left the kitchen in her sleep. On one occasion, she tried to drink from a bottle of ammonia-based cleaning fluid, only failing because the struggle to open the safety lid succeeded in waking her up.

\* \* \*

Sleepwalking can even be deadly. Without doubt, there are people who have committed murder in a sleepwalk and gone to prison for it. 'I firmly believe any such conviction is unlawful,' wrote Bill Dement in *The Promise of Sleep*. The monumental challenge is trying to establish, after a tragic event, whether someone was sleepwalking or not.

In the 1840s, Rufus Choate, a sharp Bostonian attorney, became the first person in America to use sleepwalking as a successful defence against the charge of murder. His client, Albert Tirrell, had fallen in love with a prostitute called Maria Bickford and left his wife and two children for her. When she was found with her throat slit from ear to ear and the room set on fire, Tirrell was the prime suspect, especially as there were several people who'd witnessed him running from the scene. But Choate argued that Tirrell had not murdered Bickford at all or if he had he'd done it in his sleep, bringing forth several witnesses who testified that Tirrell had a history of sleep walking. With direction from the judge, the jury acquitted him of both murder and arson, though he was sentenced to three years' hard labour on account of adultery and lascivious cohabitation.

Homicidal somnambulism has been used as a defence in a handful of cases since then, but the most celebrated of these, by far, is that of a young Canadian man called Ken Parks. On 23 May 1987, he was sitting up watching *Saturday Night Live* until the small hours of the morning. At around 1.30 a.m., he fell asleep on the couch. This wasn't particularly unusual, because he was something of a night owl, going to bed late and rising late. What was a surprise was waking to find himself stabbing his mother-in-law, especially as she and her husband lived some 23 km away and he'd driven to their house, navigating several sets of traffic lights along the way.

Parks did not remember getting up from the couch and putting on his shoes and jacket. He had no recollection of leaving his house and locking the door. He could not recall getting in his car or anything of the 15-minute journey. He had no memory of entering his parents-in-law's house, strangling his father-in-law into an unconscious state, fracturing his mother-in-law's skull by beating her with a blunt instrument and then stabbing her five times in the chest and neck. Covered in blood and with several of his own tendons severed, Parks did recall fragments of the journey to the local police station. 'I think I have killed some people ... my hands,' he told the on-duty staff.

Sad, remorseful and perplexed, Parks cooperated fully with police, physicians, psychologists and lawyers. In the absence of a compelling motive (he had always had a good relationship with his mother-in-law, who referred to him as her 'gentle giant'), Parks' lawyer called on a sleep specialist to assess the accused. Roger Broughton, then at the University of Ottawa in Canada, was 'initially skeptical that such events could have occurred during an episode of somnambulism', but the more he investigated, the more it began to look like this is what had happened.

Parks had been a severe bed-wetter until he was about 12 years old, was a chronic sleep talker and an occasional sleepwalker since early childhood. Interviews with Parks' wider family revealed that several of his relatives also experienced these exact same parasomnias. Parks had felt no pain from his own wounds until after he arrived at the police station, just one of several details that are characteristic of sleepwalking. In prison, awaiting trial, a couple of Parks' cellmates reported that he sat up in bed, eyes open and mumbling. An overnight test revealed a highly irregular architecture to his sleep. With Broughton's testimony, Parks' lawyer successfully argued that this had been a tragic case of automatism. The jury agreed, resulting in a complete and unqualified acquittal.

\* \* \*

REM sleep behaviour disorder (RBD) can be similarly violent. As we've learned, REM is normally accompanied by a complete loss of muscle tone around the body. In RBD, this muscle blocking does not work, with the upshot that you act out your dreams. It's not the same as sleepwalking, a parasomnia that occurs during the deepest, least rousable stages of sleep. But RBD can be very dangerous nonetheless.

Miguel de Cervantes appears to have unwittingly written an episode of RBD into *Don Quixote*, where the eponymous

hero clearly acts out a dream. 'Stay, robber, scoundrel, pol-
troon; I have you at last; and your scimitar shall not save you!'
he shouts, as Sancho Panza enters to find his master beside
his bed with a blanket wound round his left arm and a sword
in his right, slashing at wineskins rather than the giant of his
dreams.

But acting out dreams is rare, and was only formally
described in 1986. The first patient mentioned in the paper
was a 67-year-old man with a habit of punching and kicking
his wife during his sleep, falling out of bed, staggering about
the room and crashing into objects. It was clearly different
from sleepwalking. In fact, he was dreaming. In the most vivid
episode he recounted to his doctor, he was a halfback in an
American football team. Receiving the ball from his quarter-
back, he had to dodge a tackle. As is allowed in the rules,
he gave the incomer his shoulder to bounce him out of the
way, only to wake up in his bedroom. 'I had knocked lamps,
mirrors, and everything off the dresser, hit my head against the
wall and my knee against the dresser.' On another occasion,
he was dreaming he was riding a motorbike when another
biker came up alongside and tried to ram him off the highway.
He kicked out at the offending vehicle, which is when his
wife woke him up in considerable alarm, shouting, 'What in
heavens are you doing to me?'

In 2008, Brian Thomas and his wife Christine took their
camper van on holiday to Wales. As they settled down for the
night in the seaside village of Aberporth, they were kept awake
by an unruly bunch of 'boy racers' performing wheelspins and
handbrake turns. The couple decided to move the van to a different part of town, where they comfortably fell asleep. In the
night, Brian dreamed that the youths had broken into the van,
and in his sleep he gallantly fought them off. Unfortunately,

he ended up strangling his wife to death. Based on his history of sleepwalking, the medications he was taking at the time, his immediate panic-stricken call to the police and other testimony, it was judged that he'd been asleep. It is unclear precisely what state of parasomnia he'd been in, but because he'd been dreaming, a bout of RBD is certainly possible.

Following Michel Jouvet's discovery that REM originates in the pons, subsequent generations of neurologists have been busy working out the chain of events that results in the floppy muscles that normally accompany this state. This circuitry is key to understanding both RBD and cataplexy. When the REM neurons are activated during sleep, a neurological pulse travels upwards, filtered stained-glass-style through the cortex to fashion dreams. In parallel, the REM neurons also fire excitatory signals downwards into the medulla at the very top of the spine. These medullary cells, when stimulated, shoot a strong inhibitory signal to the motor neurons nearby, effectively preventing any muscle-related signals travelling from the brain to the rest of the body. The result is muscle atonia, says Pierre-Hervé Luppi, a neurologist at the University of Lyon, who trained under Jouvet and whose group has teased apart this pathway.

If there's damage to any of the cells in the line of communication between the pons and the medulla, then the REM signal may happily forge dreams but without the muscle block. Many cases of RBD are thought to result from neurodegeneration of the paralysing pathway. For a start, the prevalence of RBD increases with age. In post-mortem investigations, RBD brains reveal signs of degeneration of the brainstem. In many cases, someone with RBD will also go on to develop Parkinson's Disease or another neurodegenerative disorder.

Acting out dreams happens to be very common in

narcolepsy, with around one in three narcoleptics displaying RBD-like behaviour, though these events are often less severe than in full-on RBD, they affect men and women equally (where RBD is almost exclusively a male phenomenon) and there is no clear correlation between RBD-like activity in narcolepsy and the onset of neurodegenerative disease. It's much more likely that the loss of the hypocretins is behind the enactment of dreams in narcolepsy.

In the case of cataplexy, it's almost the reverse of RBD, a flood of emotions activating the REM paralysis but without creating dreams. Luppi is close to isolating exactly where the errant cataplectic signals wash up, in the pons or in the medulla. The answer has practical implications. 'If you understand the mechanism you can counteract it more easily,' he says, suggesting that the work could lead to drugs that offer far more exacting control over cataplexy than anything currently on the market.

\* \* \*

If people can talk, walk and even drive in their sleep, then it seems as though anything's possible, even sex. In the 1980s, South African psychiatrist Colin Shapiro was being interviewed by a journalist about his sleep research. When the interview wound up, the 38-year-old woman – known in the case literature as AK – told him she had another query, one of a highly personal nature. She was prone to masturbate in her sleep. When her husband woke her up 'she was always unaware of her behaviour and very embarrassed,' wrote Shapiro. In the years that followed, Shapiro began to think this kind of sexual activity during non-REM sleep might not be a one-off but a new kind of parasomnia. He suggested the name 'sexsomnia'.

There are several kinds of atypical sexual behaviour that can occur during sleep. First, there are activities like moaning that occur on a nightly basis, annoying to the bed partner but nothing more. Second, there are those behaviours that are annoying to the bed partner and harmful to the sexsomniac, one man suffering 'regular bruising of the penis and soreness of the groin'. Third, and much more alarming, are those occurrences that are almost indistinguishable from rape.

It's not surprising perhaps that more people don't tell their doctors about this. As a result, sexsomnia could be a whole lot more common than we realise. Michael Mangan, a psychologist currently at the University of New Hampshire in the US, has tried to look into this using a web-based survey. There are obvious problems with this approach. 'I sleep in a barn and had sex with a chicken,' reported one correspondent. But after removing these pointless trolls, the study had collected more than 200 plausible cases of sexsomnia.

This is something that needs to be taken very seriously. It can put a strain on relationships. 'I'm afraid of my husband when he's asleep. I fear for my daughter,' one woman wrote in an online survey, one of many first-hand accounts that Mangan brought together in a short book, *Sleepsex: Uncovered.* 'I try to be understanding and to feel that this is not his fault, but I'm still angry and hurt.' Her husband, she reported, felt wracked with guilt and had lost all self-esteem. 'I love him, but this has got to stop!' wrote another woman. 'I don't know how much longer I can take the fear of going to bed with my Jeckyll-and-Hyde [sic] husband.' One man who got in touch with Mangan was crying for help, fearing the worst for his relationship with his girlfriend. 'This is causing a huge rift between us, and I am afraid I may lose her unless I do something that stops the sleep sex.'

Just as sleepwalking has been successfully used as a defence against the charge of murder, so sexsomnia has been used as a defence against rape, and Shapiro has acted as an expert witness. In 2003, for instance, Jan Luedecke was accused of sexual assault of a woman at a house party in Toronto. She'd fallen asleep on a couch and woke to find him having sex with her. She pushed him off and, according to Luedecke's testimony, it was only when he hit the floor that he became aware of what he was doing. As with sleepwalking, it's impossible to be certain exactly what happened after an event like this, so the decision the jury reaches will rest heavily on which legal team makes the best argument.

In this particular case, Shapiro was able to establish that Luedecke had a history, with four previous girlfriends testifying he had had sex with them while he was asleep. On this evidence, sexsomnia was judged to be a plausible defence and Luedecke walked free. The woman he'd assaulted broke down in court.

There is no question that it's possible to kill or assault someone whilst in the throes of sleep terror, sleepwalking, RBD or sexsomnia. It is therefore legitimate that sleep disorders could form part of a legal defence. The concern is that this defence is abused, someone with a history of parasomnias knowingly committing a crime but then using their condition to evade conviction.

In comparison to RBD and sexsomnia, sleep paralysis is a whole lot more common. It might not often result in harm, but it is terrifying nonetheless.

# 9

# Ghosts and demons

*'In the night time, when she was composing*
*herself to sleep, sometimes she believed the*
*devil lay upon her and held her down.'*

Isbrand van Diemerbroeck

I wake to find my body gripped by some kind of impossible paralysis. But I am more alarmed by the icy wave of terror that has just rushed over my supine frame and drained into every pore. There is someone in the room with me.

In the movies, it's not uncommon to find the supernatural action hero closing his eyes to summon up his strength before unleashing his focused, air-rippling life-force upon his nemesis. I try to do something similar, willing myself to sit up, to shout, to lunge at the imposter. Only nothing happens, my loose, useless muscles mocking the wild panic inside my head.

To a fly-on-the-wall observer, there would be little to distinguish my body from that of a corpse. I know this because in my mind's eye, I see the world as if I am that fly. I watch myself lying on my bed, flat on my back, legs straight, arms still. There is no detectable rise and fall of my chest. There is no sight or sound of my panicked heart thumping out its silent call for help. The only movement is in the occasional, spasmodic twitch of an eyelid. But these can be tricky to spot. It looks like I am dead.

In fact, I am more alive and more alert than ever, locked in a desperate struggle to confront the hooded figure standing just inside my bedroom door. I didn't think it was possible to be more frightened, but he turns towards me, sharp eyes in a bony face fixing me from inside his hood. Then he starts to advance.

I was silently shouting before; now I am silently screaming. The thing – the personification of purest evil – stops at the bedside and theatrically draws an axe from beneath his cloak. He clasps it in two hands and slowly raises it until it is poised, pendulously, above me. My inaudible outrage collapses into inaudible whimpering and I give in to the inevitable. The blade falls, accelerating under its own weight to bury itself in my chest. I feel my skin split open, tissue divide, ribs shatter, a searing pain and the tickling sensation as the blood trickles down my sides. How well the sternum accommodates a hand-axe, I think.

At last, I twitch a limb. I manage an incoherent slur, but at least it's audible. I struggle into a sitting position and make a feeble lunge for my attacker. But there is no one there. No axe, no blood. The only evidence I have that anything untoward has happened is my thundering heart, sweaty palms and sense of dread.

I have been here before. Many times. In fact, I know that as soon as I calm down enough to drift off to sleep, the axe murderer will be back. Strangely though, no amount of repetition can diminish the terror. By the time I have been butchered five times in the space of 20 minutes, I am an exhausted, trembling wreck. I desperately need to rest, but I am too afraid to sleep.

\* \* \*

It is some consolation to learn that this kind of experience is pretty common; it's reckoned that around four in every ten people probably experience something like this at least once in their lifetime. It's also clear that humans have been describing this phenomenon for millennia. Almost 2,000 years ago, the renowned Greek physician Galen coined the term 'ephialtes', literally 'the one who leaps upon'. The Romans came up with 'incubus' from the Latin verb 'incubare', 'to lie down on'. In fact, there is a word with a similar meaning in just about every language you care to look. In Arabic, there is the 'Jathoom', translating as 'what sits heavily on something'. The Japanese have 'kanashibari', a sense of being 'bound by metal'. In Estonia, 'Luupainaja' translates as 'the one who presses your bones'. The Germans have 'Nachtmahr', the French 'cauchemar', the Dutch 'nachtmerrie', the Polish 'zmora', the Russian 'kikimora', the Icelandic 'martröd', the English the 'mare', and so on. All these words in all of these different languages embrace the same feeling the Greeks and Romans were trying to describe with the ephialtes and incubus: the presence of some kind of spirit crushing its dormant victim. Indeed, in its original incarnation, this is what 'nightmare' meant and it is only relatively recently that the word has come to refer to 'bad dreams' in general.

In a world before science, the most plausible explanation for the original nightmare lay in the supernatural. In some cultures, the words make explicit mention of what the presence might be. Sometimes, it's ghosts, as in the Chinese 'meng yan' (literally 'ghost oppression') or the Thai Phi Am ('ghost covered'). Often it's some kind of deity, like the Greek god Pan or the Gaulish devil Duses. The Germans talk about being squeezed by elves (Alpdrücken) or witches (Hexendrücken), the Hungarians refer to 'witch pressure' (boszorkány nyomás) and the English to being 'hagrod' or 'hagrid' (literally being ridden by a hag).

These references clearly bundle up two symptoms that sleep specialists like to keep separate: sleep paralysis and hallucinations that occur on the way into or out of sleep. In my own experience, these two phenomena almost always occur together. But it is true that sleep paralysis can occur without the overt hallucinations. Indeed, one of the first times I experienced either of these symptoms, it was the paralysis that stood out.

\* \* \*

It was in the summer of 1994, I had just turned 21 and I was on that trip round India. On 31 August, I checked myself in to a very basic room in a very basic hostel in the Rhajastani town of Satna. I had come to see the world-famous Hindu and Jain temples at Khajuraho. But tired by the long bus journey from Agra (and because I was already showing signs of pathological sleepiness), I dropped my backpack, flipped on the overhead fan and lay down for an afternoon nap.

After all these years, my recollection of what happened next is limited, but thankfully I still had that delightful youthful obsession with keeping a diary. Reading the entry for 31 August, I marvel at the clinical nature of this record, written just moments after coming out of what, I now know, was my first experience of sleep paralysis:

Just having an afternoon nap, the strangest abnormality came over me. My mind awoke but my body was still sleeping. I could feel myself suspended as if caught between two worlds. It felt, I imagined, like inhaling one's first breath of water when drowning. Although it sounds like a dream coming close to consciousness, it was very clearly no longer a dream. The fan above me was making its noise, I could feel the air, my eyes were open for much of the time but falling closed and providing a blurred and sporadically black picture. My breathing was at a completely different rate – slower and I had to work at it. At one point I relaxed, which felt much better, but the breathing stopped. As I struggled to resume it, my lungs seemed to whir or gurgle upon inhalation, an effect that disappeared and could not be reproduced as soon as I regained control; there was no loose phlegm, which might have accounted for it.

The experience was very frightening and was one, I sensed, that was intimately 'linked with death'. In fact, it was the feeling that I had nearly died and might be about to do so that compelled me to record the paralysis in such detail. It was not so much the prospect of death itself, more a concern that there would be no coherent explanation for friends and family. 'Hence I write this before slipping back into a reverie,' I scrawled before dropping off again into fitful sleep. My diary entry for the next day is short. Its tone is surprised: 'I'm still alive.'

The historical, artistic and literary record is littered with clear-cut examples of sleep paralysis combined with hallucinations. One of the first doctors to come at the phenomenon from a scientific perspective was an English physician called John Bond, who wrote *An Essay on the Incubus, or Night-Mare*, way back in 1753. In the preface, he came clean about his motive for writing: 'Being much afflicted with the Nightmare, self-preservation made me particularly inquisitive about it.' Bond was convinced that there was a scientific explanation for his terrifying dreams. More than 50 years later, in 1816, English physician John Waller once more urged the medical profession to take the nightmare more seriously (he was talking about sleep paralysis hallucinations here, rather than just bad dreams): 'Its real nature has never been satisfactorily explained, nor has it by any means met with that attention from modern physicians which it merits.'

The French responded, with psychiatrist Jules-Gabriel-François Baillarger studying the phenomenon in some detail. In 1846, he argued that hallucinations fell into one of two categories: physical hallucinations that are a product of the imagination alone and psychosensory hallucinations that combine both the imagination and the senses. It wasn't a bad hypothesis, for we now know that the hallucinations experienced during sleep paralysis involve plenty of interaction between those parts of the brain that handle sensory information and the higher 'thinking' part of the brain known as the cortex. Baillarger's work inspired much head-scratching amongst French scientists, with Alfred Maury (who also had first-hand experience of these trips) describing them as 'hypnagogiques', hallucinations that occurred, literally, 'on the way into sleep'. Just to confuse things, a self-confessed psychic by the name of Frederic Myers added another adjective – hypnopompic – to refer to hallucinations that occurred on the way out of sleep. But this distinction is most commonly collapsed, and these hyper-real experiences in the borderlands at either end of sleep are generally referred to as hypnagogic hallucinations or, more simply, hypnagogia.

The inability to move, which is at the heart of sleep paralysis and hypnagogia, is a strong indication that REM, with its by-now-familiar muscle atonia, is probably involved. In the 1980s and 1990s, a team of Japanese researchers hit upon the idea of disrupting the sleep of volunteers to increase the incidence of REM occurring shortly after falling asleep (as is commonplace in narcolepsy), hoping that this would bring on bouts of sleep paralysis. It worked.

The REM state also explains why people in the throes of sleep paralysis so often report some kind of hallucination. In fact, all five of the Japanese subjects who became aware of a REM-induced paralysis reported some other strange experience, most of them intensely alarming. Like the female subject known only as AK, who felt her feet were paralysed and awoke to see 'the strange face of a man on the wall of the bedroom'. Or KK, who felt her body tighten and had trouble breathing. Or TI, who reported strange sounds 'like a cassette tape winding quickly'.

The most common hallucination is the sensation of a presence, reported by around 60 per cent of people suffering sleep paralysis. In most of these cases, the person will actually see something, often vague, sometimes animal, occasionally distinctly human. But the hallucinations come in plenty of other guises too.

Sounds are common. In the main, these are abstract and only vaguely menacing noises – buzzing, wheezing, rustling, growling. But sometimes, they can have a narrative logic that makes them very scary indeed, like a key turning in the lock, then a door opening and closing, footsteps approaching. In *Wide Awake and Dreaming*, fellow narcoleptic Julie Flygare recounts several such episodes in which she awoke, convinced that someone had broken into her flat. I know first-hand how unsettling this can be.

I am also familiar with the sense of being touched, the bedcovers sliding on and off my body, and the fear. I have never experienced smells, but some people do: not cut grass and roses but aptly repulsive odours like mould or, in one case, 'rotting flesh'.

It's important to note that these are not bad dreams – they are categorically different. The most important distinction is that these hallucinations are accompanied by a clear sense of having woken, which explains why they always occur in your immediate, real-world surroundings. If they do not, they do not qualify. An experience, for instance, in which you are being chased around a castle by a monster falls into the more benign category of 'bad dream'. As scared as you might feel, it is nothing like the visceral terror brought on by these waking paralyses.

As the Japanese study indicated, sleep paralysis can be brought on by disrupting sleep, which is why it's pretty common in insomnia. During a particularly brutal phase of wakefulness, for instance, filmmaker and insomniac Carla MacKinnon began to experience regular episodes of sleep paralysis. 'My eyes were shut but I knew for sure that someone was standing above me, watching me. The mental image was crystal clear, clearer than actual vision, as strong as the unmistakable sensation of being stared at. I felt my whole body shake like mad under this enormous, trembling pressure as I tried to move under a fog of tingling static. I was physically straining to wake up – my will battling with my inert body: 'Move! Do something! The unconscious has taken me hostage!' Profoundly troubled, Carla began to look into the phenomenon and ended up making an award-winning short film. *The Devil in the Room* captures the condition perfectly and includes a disturbing sequence of a corpse-like character pressing himself down upon the helpless female victim, his skeletal hands crawling over her face. For women who experience sleep paralysis, such sexual predators are alarmingly common.

Yet although doctors have been talking about sleep paralysis and hypnagogic hallucinations for at least 150 years, these phenomena have remained at the margins of mainstream neuroscience. This is, in large part, simply because studying the sleeping brain is incredibly hard to do. But there is probably another reason why sleep paralysis and hypnagogia have not received serious attention until recently. For most of human history, these frightening experiences have been understood by recourse to the supernatural, variously interpreted as ghosts, demons, vampires and witches, and more recently subject to the warped psychoanalytical logic of Sigmund Freud. It is to these bogus but thrilling interpretations that we now turn.

* * *

Carol was staying with friends and had opted to go to bed early. She was facing the closed bedroom door, when it opened to reveal the hall light. 'There was this bright shimmering substance ... this very vaporous looking thing, and as soon as I saw it, I was just stiff. And I couldn't move. I was just scared stiff,' she told sociologist David Hufford, who carried out one of the first systematic studies of sleep paralysis in his book *The Terror That Comes in the Night*. The object, whatever it was, began to float towards her, breathing heavily all the while as it passed around the bed. Her spine began to tingle with a strange intensity and the image of a knife flashed into her mind. She had the strong feeling she was about to be stabbed in the back. 'I was so scared I couldn't scream. I couldn't get up and leave the room. I was just paralyzed!' She began to feel something pressing down on her, pushing her body into the bed and the bed down into the floor. Eventually, the presence moved back to the foot of the bed, then drifted back out of the door, closing it behind it. 'This was the first time I ever really realised I'd come in contact with a ghost,' she said.

Kris Jackson, a young Glaswegian narcoleptic with fabulously lucid dreams, often finds a ghost-like face peeping in at his bedroom door. Kris is able to dispel these spectres by staring at them. 'They would see you were looking and they would hide,' he says. If he isn't able to because of the position he's lying in, the intruders can drift over to his bedside and lean in over his body, as if trying to see his face without themselves being seen. As far as Kris is concerned, he *is* seeing ghosts. 'These are real things that only certain people can see.'

For people of faith, hypnagogia will often take the form of the devil. One of the earliest such accounts dates back some 350 years, when Dutch physician Isbrand van Diemerbroeck recorded the testimony of a 50-year-old woman. 'She could hardly speak or breathe, and when she endeavoured to throw off the burthen, she was not able to stir her members,' he wrote. As in all other accounts of this kind of experience, she managed to escape, but only 'with great difficulty', often aided by her husband giving her a prod.

In another touch of the Satans, a Mormon missionary was lying in bed on a warm and muggy evening when he suddenly felt something pressing on his feet 'so hard it nearly broke them off'. Then, whatever it was began to crawl up his legs. 'It was such terrible pressure.' Suddenly, it took the form of some kind of beast and snarled, its fire-filled eyes about the diameter of tennis balls. It was then he knew it was a demon intent on disrupting his missionary work. The man 'used his priesthood and rebuked it in the name of the Father' and the demonic figure recoiled. As it jumped off the bed and retreated, the man could distinctly hear the rustling of feathers.

Some of these individuals have actually sought exorcism. As

I'm not a religious person, I can't say this ever occurred to me. But I know the deep sense of evil and I can see how – for some people – an exorcist might seem just the ticket.

Sleep paralysis has also fed into the world of witchcraft, often with fatal consequences. Take, for instance, the testimony of Joan Jorden, a woman living in the small English village of Stradbroke in Suffolk at the end of the sixteenth century. At a distance of more than 400 years, it's impossible to be sure what happened between this woman and fellow villager Olive Barthram, but it would be safe to say that there was a falling out, for Jorden went before a magistrate in 1599 to testify that Barthram was a witch. What was her evidence? Nocturnal visitors sent to torment her, allegedly by Barthram. The core features of Jorden's account – being situated in her bedroom, the paralysis, the inability to speak, an evil presence, the sense of pressure, the invasive terror – should, by now, be only too familiar.

The first thing Jorden was aware of was a scraping of the walls, then knocking, then a rustling sound reminiscent of bulrushes waving gently at the margins of a river. Then she saw him, Barthram's black cat Gyles. 'He clapped the maid on the cheeks about half score times as to wake her … kissed her three or four times, and slavered on her, and lying on her breast he pressed her so sore that she could not speak, at other times he held her hands that she could not stir, and restrained her voice that she could not answer.' On the basis of this testimony, Barthram was convicted of witchcraft and strung up by the neck.

There is another story from the Salem witch trials of 1692, where Susan Martin faced a volley of similar accusations. One man testified that Martin had threatened him with death, whereupon he'd been visited by a cat-like figure at night, which 'took fast hold of his throat, lay on him a considerable while, and almost killed him'. Another witness had seen a nocturnal vision of Martin herself. 'She took hold of this deponent's feet, and drawing his body up into a heap, she lay upon him near two hours; in all which time he could neither speak nor stir.' Martin ended up hanging from the gallows.

These waking hallucinations may also have inspired stories of vampires. There is, for instance, a story of a 'Spectrum' that terrified an entire village in Silesia way back in 1591. Some people reported that 'it would strike, pull or press, lying heavy upon them like an Ephialtes'. Others felt it would 'stand by their bed-sides, sometimes cast itself upon the midst of their beds, would lie close to them, would miserably suffocate them'. The presence, the villagers felt, 'took the exact shape and habit' of a man who had recently committed suicide, and figuring that he was not properly dead, they got permission to dig up his body. When they did so, they found though it had 'lain in the ground near eight months' it had not decayed at all, a tell-tale sign that they were dealing with a vampire. Once they'd cut out and incinerated the dead man's heart, the spectre came no more.

It may have been this tradition that inspired Bram Stoker to describe one of Dracula's more terrifying visitations in his 1897 gothic horror classic. 'There was in the room the same thin white mist that I had noticed before,' Mina Harker explains to Professor Abraham Van Helsing. 'I felt the same vague terror which had come to me before and the same sense of some presence ... Beside the bed, as if it had stepped out of the mist ... stood a tall, thin man, all in black', a figure that tallied closely with the description of the vampire that others had reported. 'The waxen face; the high aqualine nose, on which the light fell in a thin white line; the parted red lips, with the sharp white teeth showing between; and the red eyes that I had seemed to see in the sunset on the windows of St Mary's Church at Whitby … For an instant my heart stood still, and I would have screamed out, only that I was paralysed.'

Then there's Freud. For just when medical men like Bond, Waller, Baillarger and Maury seemed to be making headway with the ghosts, demons and vampires and putting hypnagogia on a serious scientific footing, in walked Sigmund and his couch. He and his influential followers, notably the British neurologist and psychoanalyst Ernest Jones, argued that hypnagogic hallucinations were a result of 'some repressed component of the psycho-sexual instinct' and 'reactivation of the normal incest wishes of infancy'. Hmm.

This quackish legacy is very much in evidence in the failure, in 1982, to diagnose Michelle Hicks' narcolepsy. Michelle's parents – worried about their seven-year-old daughter's daytime sleepiness and chilling nightmares – found themselves referred to a psychiatrist at the Whittington Hospital in north London. 'It would appear that her sleeping during the day can be put down to the fact that she is not actually sleeping during the night,' he wrote. It seems the psychiatrist knew nothing about narcolepsy or the hypnagogic hallucinations that frequently accompany it. So he made a series of Freudian observations that he felt might account for Michelle's unusual sleep and dreams.

'For some little time she has been experiencing nightmares and has then been going into the parental bed.' Michelle's father, he noted with interest, had displayed similar behaviour, sleeping in his own parents' bed until he was nine. Michelle's mother had lost her father at the age of six and had never got over it. Michelle's father had been in the army, stationed in Northern Ireland for a time, creating 'anxiety about whether or not he might be killed there' and reawakening 'all the feelings Mrs. Hicks had about the loss of her own father'. But it was when Michelle's mother mentioned that she wanted a second child that the ghost of Freud really floated to the fore. 'A further function of Michelle's coming into their bed at night is to keep them apart,' he attested, 'giving them no privacy and thus ensuring that no other children do arrive'. It turns out there was a simpler explanation, though not one that Michelle would get until she was 35.

\* \* \*

The list of artists and writers who have reproduced hypnagogic hallucinations in their work is nothing short of extraordinary. Easily the best-known example is *The Nightmare*, created by the Swiss painter Henry Fuseli in 1781, which depicts a woman lying on her back, muscles limp, with a demonic imp crouching on her chest. In the background, a horse has nuzzled its way through a curtain, its milky white eyeballs overseeing the intrusion. But for Brooklyn psychiatrist Jerome Schneck, writing in the *Journal of the American Medical Association* in the late 1960s, it was the 'demoniac character squatting on the abdomen and chest of the supine woman' that suggested *The Nightmare* is a portrayal of sleep paralysis with its accompanying hypnagogia.

A few years later, the Enlightenment physician and polymath Erasmus Darwin (Charles' grandfather) published *The Botanic Garden*, a collection of poems about nature in which he put words to Fuseli's famous painting:

Back o'er her pillow sinks her blushing head,
Her snow-white limbs hang helpless from the bed;

While with quick sighs, and suffocative breath,
Her interrupted heart-pulse swims in death …
And stern-eyed Murderer, with his knife behind,
In dread succession agonize her mind …
In vain to scream with quivering lips she tries,
And strains in palsied lids her tremulous eyes; …
On her fair bosom sits the Demon-Ape
Erect, and balances his bloated shape;
Rolls in their marble orbs his Gorgon-eyes,
And drinks with leathern ears her tender cries.

Buoyed by the identification of sleep paralysis in *The Nightmare*, psychiatrist Schneck had a look for other artistic references to the phenomenon. He found them aplenty in the work of French novelist Guy de Maupassant, notably in his short story 'Le Horla' published in 1887, in which a bourgeois man is terrorised by some supernatural being at first during sleep. The first clear-cut description of a hypnagogic hallucination occurs early on in the story, after the protagonist has been asleep for two or three hours:

> I am gripped by a dream – no – a nightmare. I have the feeling that I am lying down and that I am asleep … I feel it and I know it … and I also feel that someone is approaching me, watching me, touching me, climbing onto my bed, kneeling on my chest, putting its hands around my neck and clamps … squeezes … with all its might as if to strangle me.
>
> I struggle, bound by some excruciating impotence that paralyses us in our dreams; I want to scream, – I can't; – I want to move, – I can't; – I try, with a huge effort, to breathe, to turn, to throw off this thing that is crushing and smothering me, – I can't!

And suddenly, I wake up, panicked, covered in sweat. I light a candle. I am alone.

Like a man obsessed, Schneck then found a description of sleep paralysis in F. Scott Fitzgerald's *The Beautiful and the Damned*:

> She was in a state half-way between sleeping and waking, with neither condition predominant ... and she was harassed by a desire to rid herself of a weight pressing down upon her breast. She felt that if she could cry the weight would be lifted, and forcing the lids of her eyes together she tried to raise a lump in her throat ... to no avail ... She became rigid. Someone had come to the door and was standing regarding her, very quiet except for a slight swaying motion. She could see the outline of his figure distinct against some indistinguishable light. There was no sound anywhere, only a great persuasive silence ... only this figure, swaying, swaying in the doorway, an indiscernible and subtly menacing terror, a personality filthy under its varnish, like smallpox spots under a layer of powder.

In fact, it almost seems as though writers deploy these hypnagogia as a literary device. In *The Withered Arm* by Thomas Hardy, the milkmaid Rhoda Brooks (who is made pregnant, then discarded, by a well-heeled farmer) has a vision of the farmer's nubile wife, 'with features shockingly distorted, and wrinkled as by age' and senses her 'sitting upon her chest as she lay'. In *Moby-Dick* by Herman Melville, the narrator Ishmael awakes at one point to a 'nameless, unimaginable silent form or phantom' that leaves him 'frozen with the most awful fears', adamant that if he were able to tweak his hand just 'one single inch, the horrid spell would be broken'. In *The Snows of Kili-manjaro* by Ernest Hemingway, a writer called Harry is visited by death, which 'moved in on him so its weight was all upon his chest, and while it crouched there he could not move or speak'.

This catalogue of high-profile artistic renditions – all remarkably similar in their core symptoms and all clearly recognisable as hypnagogic hallucinations – demands something of an explanation. Perhaps all of these famous artists and writers were simply running with what is undeniably a rather nifty device, with these hypnagogia being passed on from one creative mind to the next like batons in a relay. Alternatively, they were writing from experience.

When Fuseli unveiled *The Nightmare* at the Royal Academy in London in 1782, its subject-matter was so freaky that many people were quick to assume the artist must have been high on opium. The presence of a couple of cute bottles on the bedside table only added to the speculation. These trinkets, which Fuseli faithfully reproduced in several subsequent paintings on the same theme, were just the sort of props that might contain laudanum.

There is certainly evidence that opium can bring on hypnagogic hallucinations. In *Confessions of An English Opium-Eater*, for instance, Thomas De Quincey writes how the drugged-up individual will often lie 'under the weight of incubus and nightmare':

> he lies in sight of all that he would fain perform, just as a man forcibly confined to his bed by the mortal languor of a relaxing disease, who is compelled to witness injury or outrage offered to some object of his tenderest love: he curses the spells which chain him down from motion; he would

lay down his life if he might but get up and walk; but he is powerless as an infant, and cannot even attempt to rise.

Maupassant too is known to have abused opiates, but it may have been his well-documented insomnia (or the potassium bromide he was prescribed for it) that gave him first-hand experience of the hallucinations that make 'Le Horla' such a riveting read.

Drugged up or not, this catalogue of startlingly similar experiences suggests that hypnagogic hallucinations are not uncommon. Indeed, a recent review of all the studies of sleep paralysis indicates that they regularly affect around 10 per cent of the general population, though in certain groups – like students and psychiatric patients, for instance – they are more common still, affecting around one in three people.

* * *

Over the past few decades, a handful of scientists has been bold enough to take on hypnagogia, gradually drawing them out from the paranormal shadows to expose them to the glare of rationality. Allan Cheyne has devoted plenty of thought to sleep paralysis and hypnagogic hallucinations and, like Bond and Maury before him, has experienced them first-hand. His earliest attack occurred when he was just 17.

More than half a century later, Cheyne can recall the precise events with incredible clarity. He was lying on his front, trying to get to sleep when it happened. 'I just suddenly had the feeling that there was something there. Simultaneously I realised I was unable to move,' he says. He felt the bedcovers slide off and fall to the floor and something start climbing onto the bed. 'I had this impression of a kind-of-a gargoyle type of creature – the size of a small dog, I guess – on my back and chewing on my shoulder.'

In the 1990s, Cheyne – by then an established psychologist at the University of Waterloo in Ontario, Canada – began to take a professional interest in these hallucinations. His hypothesis works like this. If you become conscious during a bout of REM, the awareness that you are paralysed triggers fear and panic. The only way to make sense of these emotions is for the cortex to fill in the void, constructing a hallucination commensurate with the terror. A scenario like this, says Cheyne, would explain why the core features of this phenomenon – the paralysis, the fear, the hallucinations – are so remarkably consistent across so many different cultures. 'People are having the same kind of experience all over the world because they are all going through the same kind of state,' he says. 'Their brains have gone through the same evolutionary processes. It all falls together rather nicely.'

By creating an internet survey, Cheyne has also been able to make generalisations about hypnagogia, identifying two main types of hallucination. The first category is 'intruder hallucinations', where there is a presence (sometimes seen, sometimes not) and an urgent terror. My axe-murdering companion is clearly one of these. The second category is 'incubus hallucinations', which involve the sense of pressure, trouble breathing and pain. My experience in Satna, with my whirring chest, ticks this box.

Cheyne's dataset also indicates that sleeping on the back increases the likelihood of suffering sleep paralysis, an observation that tallies with anecdotal wisdom that dates back centuries. In Arabic culture, for instance, the remedy for nightmares is to sleep on your side. In Shakespeare's *Romeo and Juliet*, Mercutio identifies Queen Mab as the hag that comes 'when maids lie on their backs'. Dutch physician van Diemerbroeck's advice to his sleep-paralysed patient was to avoid sleeping face up. The eighteenth-century physician Bond, who noted this association too, reckoned that supine sleeping resulted in 'a stagnation of the Blood' that put pressure on the 'tender dilatable vessels of the Brain'.

This may, in fact, be part of the explanation, as blood pressure does indeed increase when lying on your back. In addition, the heart places greater pressure on the lungs, reducing the opportunity for gas exchange and the amount of oxygen reaching the brain (and other tissues). 'One very real possibility is that these factors fracture the sleep cycle, bringing individuals into the trippy REM state more often,' says Cheyne.

If sleeping on your side isn't helping, another idea is simply to relax, transforming the most common types of hallucination – the intruder or the incubus – into a third, less common but more benign category. Cheyne calls this the 'vestibular-motor hallucination' or, to put it more simply, an out-of-body experience. 'I never had the very elaborate out-of-body-experiences until I started striving for them,' he says.

I have lived with narcolepsy for more than 20 years and, until I spoke to Cheyne I had never had one of these experiences either. But under his tutelage, I have now learned how to direct sleep paralysis away from the intruder and the incubus towards this far more positive experience, which can then tip over into a lucid dream, where you have complete control over the content of the dream. In fact, people with narcolepsy, with their pathological talent for entering REM within minutes of falling asleep, are able to enter a lucid state far more readily than most.

Vestibular-motor hallucinations are interesting for another reason. They might account for some reports of alien abduction.

In favour of this idea, most accounts take place at night, when the victim is either about to fall asleep or has just done so, and many feature several of the tell-tale features of sleep paralysis, including the inability to move, the breathlessness, a presence, the intense fear and occasionally an out-of-body experience.

One female abductee described waking to find herself completely paralysed and levitating above her bed. Her heart was pounding, her breathing was shallow, she was tense and terrified. She was able to open her eyes, and when she did, there were three figures standing in the glowing light at the foot of her bed. It might have been aliens. Then again, it sounds like a pretty textbook case of sleep paralysis to me.

\* \* \*

When I began my career as a science journalist some 15 years ago, I am not sure I would have been comfortable writing about a phenomenon that has been linked for so long to the supernatural. In spite of the fact that I have first-hand experience of the horrors of sleep paralysis, I felt that any attempt at a rational explanation would be overwhelmed by non-science tales of spectres, devils, witches, vampires, Freudian wish fulfilment and aliens. But rather than being a hindrance, I now realise that these ghosts and demons are a help. For once these figures are brought together, all assembled around a very precise set of physiological conditions, it is easier to see them as the ghoulish figments that they are, each interpretation imagined by a succession of highly subjective human brains.

We are still a long way from understanding how the brain can wake yet retain elements of REM and what is going on at a neurological level when we experience a hypnagogic hallucination. But suffice to say that sleep paralysis is a very real, very common phenomenon, and one that will be explained by science, not monsters.

## 10

# Wide awake

*'The worst thing in the world is to try to sleep and not to.'*
F. Scott Fitzgerald

I was in London, at a private screening of *Meet the Sloths*, an award-winning documentary set in the world's only sloth sanctuary in Costa Rica. Since I didn't know anyone except the film's presenter Lucy Cooke and as she was already surrounded by keen faces, I introduced myself to a stranger. The conversation did not start well. Jennifer Salinas was irritable. She was monosyllabic. It was awkward. Then, just as I was about to conjure up an excuse and reverse myself out of the situation, she apologised. 'I'm sorry for being so grouchy,' she said. 'I suffer from insomnia and had a terrible night last night. I wish I were narcoleptic.'

I burst out laughing. I have spent most of my narcoleptic life wishing I were an insomniac. If there were some fantastical way to exchange sleep disorders, I think Jennifer and I might have done so there and then. It would have been fun for a few days, she thrilled by the ability to fall asleep instantaneously, me delighted to find myself suddenly unburdened by sleep. But how long would the euphoria have lasted? One week? Two? Then, like the characters in some children's fable, we would have come to regret – too late – what we'd wished for. Blinded by the deadly sin of envy we'd swapped one sleep disorder for another when what we'd both wanted was the same thing: good sleep.

People with narcolepsy spend so much time fending off sleep, it is perhaps understandable that many come to assume that sleep is a behaviour at which they excel. In fact, this is rarely the case. In Westphal's description of the bookbinder Herr Ehlert's narcolepsy, he noted his patient's 'persistent night-time sleeplessness' and that 'he spends only a very small portion of the night sleeping'. But it's only recently that disturbed nocturnal sleep has been recognised as a fifth symptom of narcolepsy (in addition to daytime sleepiness, cataplexy, sleep paralysis and hypnagogic hallucinations). It's now thought that well over half of those with narcolepsy will wake up far more than normal during the night.

So it turns out that insomniacs and narcoleptics have far more in common than meets the eye. People with insomnia have difficulty initiating or maintaining sleep (despite ample opportunity to do so). Those with narcolepsy almost never have a problem getting to sleep, but they can have tremendous difficulty maintaining it. Although insomnia is a very common sleep complaint, with around one in ten people experiencing chronic insomnia at any one time (defined as sleeplessness on at least three nights a week for at least three months), it can be just as isolating an affliction as narcolepsy. People with insomnia and narcolepsy both end up fighting against sleep during the day. The underlying reasons for this battle might be different, but the experience is remarkably similar. Finally, people with insomnia can now do as Jennifer wished and experience a temporary bout of narcolepsy, because the first drug to target the hypocretin system and reach the market was not designed to treat narcolepsy but to help with insomnia.

* * *

Consider the market for a drug that brings on narcolepsy for a short period. It's massive. In the US, the 'sleep aid' market is worth around $1.5 billion a year, but most of these drugs work by suppressing the central nervous system. 'These drugs produce numerous side effects, including reduced cognition and what many call the "hangover effect" – an overall feeling of malaise,' Peter Kim, president of Merck Research Laboratories told shareholders in 2010.

Merck had a hypocretin receptor antagonist (or blocker) in the pipeline, then known as MK-4305. This drug 'if approved, will represent the first major new approach to the treatment of insomnia in nearly 40 years,' Kim announced. The US Food and Drug Agency (FDA) approved Merck's new insomnia drug in 2014, commercialised under the brand name Belsomra. In randomised, double-blind, placebo-controlled clinical trials, patients receiving the maximum dose permitted (20 mg per day) get to sleep more quickly, have a longer stretch of unbroken sleep at the beginning of the night and improved sleep duration, to mention just a few of the positive outcomes.

Belsomra's active ingredient is a small molecule that can be absorbed from the gut into the bloodstream unaltered and sneak its way into the brain. There, it binds to both of the two hypocretin receptors, preventing the hypocretins from working their magic, effectively creating an acute case of narcolepsy. Importantly, however, the half-life of Belsomra is only around 12 hours, so by the morning its effects will be wearing off.

'What's exciting about Belsomra is that it is very selective for blocking wakefulness, so it does not affect the systems that control balance, memory and cognition,' says Paul Coleman, a medicinal chemist who works at Merck's laboratories at West Point, Philadelphia, and who has led much of the research into Belsomra. 'Narcolepsy has given us a thread we can pull on to unravel a lot about what underlies the systems that govern wakefulness and sleep,' he says. 'Wakefulness is a pretty central process for everybody, whether you are a healthy person or have narcolepsy or insomnia. It's the most exciting thing I've had a chance to work on.'

\* \* \*

It was only in 2013, after I'd performed a second sleep study, that I began to realise I might not be the confident sleeper I'd imagined myself to be. With a colourful rainbow of wires feeding my brainwaves into the EEG I passed a fitful, uncomfortably hot night at the Sleep Disorders Centre at Guys Hospital in London. In the morning I felt shattered, like I hadn't slept at all. When the consultant tasked with reporting on my brainwaves told me I'd slept for over six hours, I did not believe her at all.

The conventional wisdom is that I'd experienced so-called sleep state misperception, when the subjective description of sleep differs radically from the objectivity of the EEG.

There is the odd case of someone thinking they've slept when they haven't, like the 71-year-old woman who turned up at a sleep clinic to be treated for excessive daytime and night-time sleep. She underwent an overnight sleep study and reported sleeping well, but where she felt she'd slept for nine hours, the EEG said two, a discrepancy replicated almost exactly in a second test. When presented with the results she was incredulous. 'She did not show up for two follow-up appointments,' wrote her doctors, so they couldn't offer her any treatment.

Much more commonly, however, it's insomniacs who experience sleep state misperception, which is why it often goes by the name of paradoxical insomnia. They think it took ages to fall asleep and that they were awake for hours, when the EEG suggests otherwise. Like the 39-year-old woman, for instance, who claimed she had not slept for 13 years. Every night, she'd get into bed, close her eyes and lie there without sleeping until the morning. The doctors who reported on her case were sceptical and requested a battery of tests, including an appraisal of her mental stability and personality. When these appeared to be normal, they wired up her brain to see what was going on during the night. As usual, she claimed to have been awake all night. The EEG begged to differ, suggesting she'd slept for almost seven hours.

The doctor showed his patient the results, figuring that it would put her mind at rest. 'I don't really care what your machine says,' the woman replied. She trusted her own mind more than the computer. Could she be right?

The scoring of the EEG involves breaking it up into discrete 30-second chunks for no other reason than that's the way it's done. The first machines spooled out paper at a rate of 1 cm per second and 30 cm seemed like a manageable length for a sleep technician to work with. The patient's state is assessed in each of these 'epochs' depending on the type of brainwaves that predominates. So a 30-second period that contains at least 15 seconds of sleepy waves is scored simply as sleep. In most cases, this is probably a pretty reliable way to scrutinise sleep, because most brains enter a particular state and stay there a while. But if the brain is zipping between states (chapter 8), there could be important detail that the EEG is simply not able to show.

Nathaniel Kleitman certainly saw the transition from wake

to sleep as a gradual process, one that 'involves a succession of intermediate states, part wakefulness and part sleep in varying proportions'. The Italians have a rather lovely word for this interstitial state: 'dormiveglia', literally 'sleep-waking', yet the standard method for scoring the EEG read-out leaves no room for such subtlety. A patient is either awake or in one of the four stages of sleep (1, 2, 3 non-REM and REM). Never somewhere in between.

A recent and very clever experiment exposes this shortcoming. Volunteers rigged up to an EEG were asked to hold a small stress ball in one hand, instructed to close their eyes and breathe normally. With each inhalation, they were told to squeeze the ball, with every exhalation to relax. They continued to squeeze the ball on and off for up to two minutes after a technician studying their brainwaves would have pronounced the onset of sleep.

It was a small study, with just ten volunteers, but for two of them – that's 20 per cent of the subjects – the ball-squeezing continued on and off for a full five minutes after a technician would have judged them to have fallen asleep. Most methods of recording sleep, from watching someone's eyelids and breathing, measuring their movements with a wrist-worn Fitbit or wiring them up to an electroencephalograph, tend to see wakefulness and sleep as mutually exclusive states, says Michael Prerau, an anaesthetist at Harvard Medical School and the lead author of the study. But the fact that some people continue to squeeze a ball in spite of being clinically asleep is a strong indication that the distinction between wake and sleep is not binary but a continuum, he says.

What's more, the onset of sleep is likely to play out differently from person to person. 'There is a lot of heterogeneity that we see that is currently not captured by standard metrics,' he says. 'There's some different wacky stuff in there that we just have no means of explaining right now.'

If we had a more realistic, probability-based model of sleep onset and the transition from one stage to the next, it could help account for many of the bizarre symptoms experienced by those with sleep disorders. It might also make a bit more sense of paradoxical insomnia.

During my sleep test, where my EEG chart showed I'd slept for six hours, it also revealed that I had woken up 26 times and been awake for almost two hours. Perhaps I had experienced meaningful snatches of consciousness as I repeatedly drifted from wake to sleep, transitions that the EEG algorithm simply scored as sleep. 'It's not beyond the realms of possibility,' says Prerau.

\* \* \*

In 2015, my consultant sleep specialist Hugh Selsick asked if I'd like to attend a group session of cognitive behavioural therapy to combat my insomnia – CBT-i. I was a little uncertain. I have always considered my broken nights to be an unfortunate consequence of my lack of hypocretins, untreatable in exactly the same way as my daytime sleepiness. I had a good idea of what would be involved in cognitive behavioural therapy and was doubtful that this kind of talking cure would do much to compensate for the absence of a chemical.

Even if CBT-i could help, I had another concern. I remembered how Jennifer had spoken of narcolepsy. If I walked into a room full of bona fide insomniacs, people I imagined to have far more seriously troubled nights than my own, how would I be received? A narcoleptic attending an insomnia clinic could be seen as the height of insensitivity. I imagined myself stepping over the threshold of a country pub, the hubbub of conversation withering to pin-drop silence, all heads turning to look me up and down with suspicion. I was also mindful of the cost to the National Health Service of CBT-i (probably in the region of £500 for the course of five sessions) and the thought that I'd be taking the place of someone more deserving than me. But Dr Selsick persuaded me to give it a try.

The CBT-i clinic was held at the Royal London Hospital for Integrated Medicine on Great Ormond Street, just round the corner from the National Hospital for Neurology and Neurosurgery in Queen Square where I'd been diagnosed with narcolepsy all those years ago. As I approached the building, I noted with interest a plaque showing that it used to be the Royal London Homeopathic Hospital, rebranded in 2010 when the UK's House of Commons Science and Technology Committee judged homeopathic remedies to be 'scientifically implausible' and no better than placebos. The discovery of this relatively recent history of quackery left me braced for an onslaught of non-evidenced nonsense.

It turns out I was wrong. 'Everything we will do on this course will be evidence-based.' This was one of psychiatrist Dr David O'Regan's first claims about the Royal Hospital for Integrated Medicine's CBT-i programme. I resolved there and then to cross-check every recommendation he made against the published literature. Nerdy, I know, but important when you find yourself attending group therapy, lounging in an easy chair that could almost pass for a couch in a hospital mocked by one commentator as a 'great national embarrassment'.

I quickly recognise the same incredible variation in insomnia as occurs in narcolepsy. Famed insomniac F. Scott Fitzgerald was on the money when he observed that 'every man's insomnia is as different from his neighbour's as are their daytime hopes and aspirations,' a reflection perhaps of the fiendishly complex neurological networks that might contribute to insomnia. There is, however, one thing that many people with insomnia often have in common: they are trying to sleep for longer than they can.

The widely held belief that everyone needs eight hours' sleep is extremely unhealthy. It eclipses the fact that at the level of a population, the duration of sleep describes a normal distribution or a neat bell shape, with the vast majority of people boasting an average and sitting comfortably beneath the crown of an imaginary bell. The further from this centre-point, the fewer people there are, until at each extreme, at each lip of the bell, there is one person with the shortest sleep duration in the population and one with the longest. This variation in sleep duration reflects the reality that all our brains do things differently and some people just don't need as much sleep as others.

'It is not possible to lay down any Rule as to the Length of Time necessary for Sleeping; for as this does in a great Measure depend upon Age, Habit and other Circumstances, it ought in different Persons to be different,' wrote the very wise author of a 1772 booklet on *Directions and Observations Relative to Food, Exercise and Sleep*. He then went on to suggest that 'it seems to be agreed, that it ought not in the general to be less than six nor more than nine Hours a Day'. If you are someone that can only sleep for six hours, then striving for the average of eight will guarantee disappointment and two hours of insomnia.

One of the first steps of CBT-i is to fill in a diary to help work out how much sleep each patient actually needs. Each morning, the insomniac must note down answers to several questions. When did you get into bed? How long did it take to get to sleep? For how long were you awake in the night? When did you get out of bed? The answers to these questions allows the calculation of what's called 'sleep efficiency', which is the percentage of time that you are in bed that you actually spend sleeping. When sleep efficiency starts to fall below 90 per cent, there may be a problem.

For many who suffer from insomnia, the 'How long did it take to get to sleep?' question is highly relevant, with sleep only coming after hours of frustration. For me, this column is redundant. I am usually asleep as soon as my head hits the pillow. But I do clearly have an issue with the number of awakenings. After an hour or so of what appears to be unbroken sleep, I will wake for a moment, fall asleep, wake, drift off again, and so on throughout the night. Then, at some point, usually at some ungodly hour, it is as if my brain has become tired of this sleep-wake rally and then that's it; these early awakenings occur for more than 80 per cent of people with narcolepsy. Try as I might, I cannot get back to sleep, staring at the ceiling, twisting and turning, mulling over this and that for at least an hour, often two, sometimes what feels like three. I record these frustrating awakenings in the sleep diary pinned to my fridge and I nod at the big numbers in this column. Perhaps I really do have a place on the CBT-i course after all.

\* \* \*

The Spielman Model is one of the most useful frameworks for understanding a life with insomnia. It is also known as the 3P Model, because it identifies three factors – predisposing, precipitating and perpetuating – that play a role in chronic insomnia.

There are lots of predisposing factors. Some of these you can't do much about, such as your gender, for instance.

Women suffer insomnia more than men, with around three female insomniacs for every two male insomniacs. Insomnia also increases with age, and ethnicity may come into play. Sleeplessness tends to run in families too, with insomniacs twice as likely to report a family history of the condition than those who do not suffer. These observations suggest a tangled web of genetic and environmental factors that can nudge an individual that bit closer to a bout of insomnia. If, for instance, you are a worrier (whether by nature, by nurture or both) then sleeplessness just becomes that bit more likely.

In most cases, however, predisposing factors are not enough to cause insomnia. This is where the precipitating factors come in, a sudden shock to the predisposed system that is thought to push the brain into a state of near-constant vigilance. Most people with insomnia will identify with this.

Jennifer Salinas acknowledges a precipitating factor for one of her many bouts of insomnia. In 1999, a neo-Nazi intent on terrorising those in the gay and lesbian community entered the Admiral Duncan pub on Old Compton Street in London's Soho, well-known for being a meeting point for LGBT people. He left a bag in the pub, containing nails and a bomb. Jennifer happened to be walking along Old Compton Street that evening. When the bomb went off, it killed three people, injured 70 and traumatised many more. 'Something just switched on inside my brain,' she says. 'I was awake for days after that.'

In most cases, such insomnia is short-lived, an acute phase of sleeplessness that the brain eventually resolves, either on its own or with the judicial delivery of sedatives. But sometimes, there are perpetuating factors that transform an acute bout of insomnia into a chronic, ongoing problem. CBT-i does not promise to do much about the predisposing or precipitating factors of insomnia. But it can do something about these per-petuating factors.

There are a bunch of bad habits that can perpetuate insom-nia. Correcting these comes under the banner of sleep hygiene. Take caffeine, for instance. After one or two nights of little or no sleep, a strong coffee or two might seem like a useful way to make it through the day, but filling the body with a stimulant is unlikely to improve the coming night's sleep.

Caffeine acts through several mechanisms, but predomi-nantly on the ability of a molecule called adenosine to bind to its receptors. When the body is busy carrying out lots of chemical reactions, as it is during the daytime, the levels of adenosine – a waste product – begin to rise both in and between cells in the brain. Under normal circumstances, aden-osine would bind to highly specific adenosine receptors and effect a calming influence on the intracellular chemistry. But caffeine interferes with this pathway, bunging up the adeno-sine receptors and preventing adenosine from binding. The upshot is that the cells continue to chug along at full-tilt, and it's hard to sleep.

In the human body, caffeine has a half-life of around four hours, which means that about one eighth of the caffeine from a cup of coffee necked at breakfast will still be zinging around the bloodstream at bedtime. Consider the following, not-par-ticularly-unusual regime of caffeine intake: a mug of strong tea at 7 a.m., an espresso at 8, a Starbucks Grande to take to your desk at 9 a.m., an instant at 11 and a cup of afternoon tea. Come 10 p.m., when most people are thinking of turning in for the night, there would still be around 200 mg of caffeine in the body, roughly equivalent to that in a double espresso. For some people this might not be a problem, but there's evidence that this kind of caffeine dose can destroy sleep. In one study, where volunteers were given around 200 mg of caffeine an hour before bed, it typically took them 40 minutes longer to get to sleep than normal.

Alcohol is not recommended either. Although one of the consequences of over-indulging in alcohol is that you pass out, it has a seriously disruptive effect on the architecture of sleep. Sleep pioneer Nathaniel Kleitman was one of the first to explore the effects of alcohol on sleep when he gave his (presumably eager) volunteers a tipple in the hour before bed. He found they went to sleep quickly and appeared to sleep soundly for the first few hours. As the night wore on, however, their body temperatures began to rise and they became more restless than when they had drunk the equivalent amount of water.

A few years later, researchers in Florida asked if alcohol had an impact on the amount of REM sleep and got students to drink the equivalent of around ten units of alcohol before sending them to sleep it off in 'a sound-attenuated, temperature-controlled, darkened room'. Excluding 'periods resulting from subjects urinating, vomiting, or technical difficulties,' all but one of the volunteers got less REM sleep than they had during an average alcohol-free control night. For several, the amount of REM was slashed by more than half. Whatever the function or functions of REM, too little of this state is probably as undesirable as too much.

Nicotine is a no-no too, because it stimulates a bunch of alerting neurotransmitters to flood the brain. Around 30 per cent of smokers suffer from disturbed sleep compared to 20 per cent of non-smokers. The heavier the smoker, the shorter their sleep.

Don't sleep on a full stomach is an adage with which everyone will be familiar. Eating late can not only send mixed messages

to the suprachiasmatic nucleus and threaten to desynchchro-
nise the body clock, it can also affect sleep through indirect
means such as gastric reflux.

Exercise too close to bedtime is also a bad idea. 'It raises
the endorphins,' says Dr O'Regan. This is thought to account
for the phenomenon known as 'runner's high', a euphoric
state first observed in long-distance runners but probably also
behind the buzz that many people feel following moderate
exercise. I go away and look for evidence that this is the case
and the first paper I come to is a rather nasty experiment on
cats: injecting endorphins into their bloodstream stopped
them sleeping.

\* \* \*

All of this is interesting, but most of us attending the CBT-i
course have already implemented these basic changes. In spite
of avoiding caffeine after midday, cutting back on the booze,
giving up cigarettes, eating early and putting some distance
between exercise and bedtime, it has done nothing to break
the insomnia.

This is because the real issue, the reason why insomnia
often flips from an acute to a chronic condition, is the brain's
natural talent for spotting patterns. Lying in bed awake one
night after the next can rapidly result in a process of Pavlovian
conditioning.

In the late nineteenth century, the Russian physiologist
Ivan Pavlov was studying digestion, using dogs as a model.
He noticed that when the dogs saw the technician tasked with
feeding them they began to salivate even if he wasn't carry-
ing food. This formed the basis of a new experiment in which
Pavlov flipped on a metronome just before feeding time. It

didn't take long for the canines to respond to the ticking metronome by salivating.

In 1973, American psychologist Richard Bootzin proposed that something very similar might account for many cases of chronic insomnia. 'Many insomniacs seem to organize their entire existence around their bedroom, with television, telephone, books, and food within easy reach,' he and colleagues observed. For others, it's the first time in the day when they get a chance to think and the thinking, the worrying and the planning for the next day keeps them awake. Either way, 'bed and bedtime become cues for arousal rather than for sleep,' he proposed.

In support of the idea, Bootzin made the anecdotal observation that insomniacs will often sleep well outside their bedroom, on the sofa, in a comfy chair, away from home. Confident sleepers, by contrast, who have learned to associate their own bedrooms with good sleep, tend not to sleep as well when in a strange environment.

Bootzin proposed 'stimulus control', banishing all activities from the bedroom except for sleep. The logic is alluringly simple. If the bedroom is variously used for reading, watching TV, eating, working, worrying, cleaning, sex and occasionally sleep, then – all else being equal – the odds of sleeping are one in eight. If the bedroom is only used for sleep, then sleep becomes a certainty. Bootzin prescribed that insomniacs should only go to bed when they are sleepy and if sleep doesn't come in 15 minutes or so, they should get up and leave the bedroom. 'This is one of the most powerful techniques to combat insomnia,' says O'Regan.

The evidence that stimulus control works comes mainly from studies in the 1980s and 1990s. Combining almost 60 of these into a meta-analysis published in 1994, researchers found stimulus control is effective, reducing the time taken to fall asleep by tens of minutes and how long people subsequently lie awake.

As the desperation intensifies, the insomniac will often begin to chase sleep, going to bed a little earlier and staying in bed a touch longer, hoping beyond hope that they will get some rest. 'This is a disaster,' says O'Regan. The Austrian neurologist and Holocaust survivor Viktor Frankl put it nicely in his book *The Doctor and the Soul*, when he wrote that sleep is like 'a dove which has landed near one's hand and stays there as long as one does not pay attention to it; if one attempts to grab it, it quickly flies away'. The anxiety involved only strengthens the troubled associations that are developing about the bedroom. To make matters worse, spending more time in bed virtually guarantees they will spend more time lying awake.

This revelation lurks behind another tried-and-tested treatment for insomnia, the paradoxical notion of reducing the opportunity you have to sleep. This so-called 'sleep restriction' involves figuring out how much sleep your body needs, calculated quite simply by keeping a sleep diary for a week or two, and then sticking rigorously to that duration. When I did my own calculation, it turns out that I typically sleep for seven hours, though it's impossibly fractured over the course of the roughly nine hours I spend in bed.

'When my Irish grandmother told me to go to bed at the same time every night, it was the worst advice ever,' says O'Regan. What she should have been advocating was an anchor time, he says, an immutable time in the morning to leave the bedroom. I set my anchor time at 7 a.m., subtract the seven hours' sleep that my body seems to require to give me a 'threshold time', the earliest I am allowed to head for bed. So I must stay awake until at least midnight. Then I can go to sleep, but only if I'm tired, O'Regan says. In the words of Friedrich

Nietzsche, another victim of insomnia: 'Sleeping is no mean art: for its sake one must stay awake all day.'

Sleep restriction feels like an impossible ask for someone with narcolepsy, and as I begin to try it there are many nights that I wake up with my head on the kitchen table to find I've overshot my threshold time of midnight by an hour or more. These aberrations are not helpful for the sleep restriction project. 'Every time you nap you are stealing some of your sleep fuel from the night,' says O'Regan. This is an interesting inversion of the usual logic. Most people think of sleep as replenishing the tank for the forthcoming day. By talking of wakefulness refuelling for the night, there is an emphasis on avoiding naps. The more fuel that is accumulated, the longer the sleep will run.

I'm not the only one on the course who's finding sleep restriction hard. There is one woman, with double-dyed, blue-on-blonde hair, whose night is divided into two distinct sleeps of around two-and-a-half hours each, one in the evening and the other in the early morning. It sounds remarkably like the way many people slept before the invention of electrical lighting. As we schedule our sleep, she works out that if she wants to wake at 8 a.m. (she is quite specific about this) she must stay up all the way through to 3 a.m., forgoing the evening sleep that her body has become accustomed to. After a week of sleep restriction, I ask her if it's difficult staying awake until dawn. She answers in an affirmative monosyllable.

Another woman is going through the menopause, a radical physiological change that quite evidently has an impact on sleep. In the Ohio Midlife Women's Study, well over half of all those canvassed reported trouble sleeping. The Wisconsin Sleep Cohort Study, which we encountered already in the context of sleep-disordered breathing, found that peri- and post-menopausal women are twice as likely as pre-menopausal women to be dissatisfied with their sleep. The most obvious complaint is difficulty falling asleep, but many menopausal women also suffer from hot flashes that tend to interfere with sleep architecture during the first few hours of the night. This fracturing of menopausal sleep may help to account for the fatigue that many menopausal women experience during the daytime, and may be why this woman often falls asleep in front of the television before she has made it to her threshold time.

Yet in spite of the difficulty I have staying awake until midnight and the battle I have to get out of bed at 7 a.m. every morning, come weekday or weekend, I soon notice that my brain gets into a groove and I'm making more efficient use of my time in bed.

\* \* \*

In addition to teaching us about stimulus control and sleep restriction, O'Regan talks us through a further ten methods that have been shown to help those suffering from chronic insomnia. At the end of week three, for instance, he shows us a technique known as progressive muscular relaxation. It's the last day of November and past 5 p.m., so when he flicks the light switch to the room we find ourselves sitting in semi-darkness. O'Regan returns to the centre of the room, pulls up a chair and sits, sliding down in the seat until his body is stretched out. 'Get comfortable,' he tells us. 'Now close your eyes.'

With his lilting delivery, O'Regan instructs us to contract, hold, then relax muscle after muscle. He starts with the head. 'Raise your eyebrows as high as you can, so your forehead is wrinkled up.' He counts to five. 'And relax,' he sighs. 'Now screw up your eyes and nose.' 1 ... 2 ... 3 ... 4 ... 5. Relax. 'Push your lips forward, clench your teeth and press your tongue to the roof of your mouth.' A dozen people pout hard into the gloom. 'And relax.' We move on to the shoulders, the biceps, the triceps and at this point, my consciousness starts to drift.

'And sit up.' I come to immediately and shuffle my bottom back in the chair. I have been both following and not follow-ing the instructions. I have a feeling that the last order was to point the toes, but I don't think I took part. 'How did that feel?' asks O'Regan. There is nodding and a murmur of approval. 'I fell asleep,' I say, before I consider how annoying this must sound in this company.

The idea of giving yourself a gentle flexing of muscles to relax your body and then mind has been around for almost 100 years, developed by American psychiatrist Edmund Jacob-son in the early 1920s. Jacobson wrote lots of books on the technique and knew that it had helped lots of his patients, but it was Richard Bootzin – he of the 'stimulus control' approach – who subjected progressive muscular relaxation to proper experimental scrutiny in 1974. He placed an ad in a Chicago newspaper, asking for insomniacs to volunteer themselves for an experiment. Some of the subjects were taught a version of Jacobson's technique. Others were told to give themselves time in the day to relax but were given no specific methods. A third group found themselves in a control, patiently waiting for a treatment programme that would never begin.

After four weeks, those practicing progressive muscular relaxation had almost halved the time it took them to fall asleep from over two hours to around one hour. At a six-month follow-up appointment, the situation was even better, with the experimental group getting off to sleep within just 40 minutes of getting into bed. Still a long time, but a lot better than before the trial. All subsequent investigations of the efficacy of progressive muscular relaxation have shown something similar.

At the end of the class, Dr O'Regan nips out of the room and returns with a stack of CDs that describe the sequence of muscular contractions we need to learn. At home that night, long after the rest of my family has gone to bed and while I'm waiting for my midnight threshold to come around, I slip the disc into my computer. It seems like a good moment to sear the PMR method into my brain.

The CD whirs into action, iTunes comes alive and I see that the 'Artist' on the soundtrack is none other than my consultant, Dr Selsick. His soft South African intonation is instantly soothing. 'Welcome to the progressive muscular relaxation programme,' he says a little too slowly. 'Close your eyes and get as comfortable as possible.' I lean back in my chair and uncross my arms and my legs, as instructed. 'I am going to help you achieve a deeper level of relaxation,' he says. 'You will not lose consciousness.' I do, of course, at some point after I've hunched my shoulders up to my ears, and before Dr Selsick reaches the next set of muscles.

I also lose consciousness during another technique that Dr O'Regan imparts. It is called paradoxical intention. In 1939, before the Nazis deported Viktor Frankl to the Theresienstadt Ghetto, he had already used paradoxical intention in his medical practice. It is based on two key ideas, that 'fear brings about that which one is afraid of' and that 'hyper-intention makes impossible what one wishes'. In short, 'The phobic patient is invited to intend, even if only for a moment, precisely that which he fears,' wrote Frankl in *Man's Search for*

*Meaning*, first published in 1959. 'By this treatment, the wind is taken out of the sails of the anxiety.'

Russian author Fyodor Dostoevsky made the same observation. 'Try to pose yourself this task: not to think of a polar bear, and you will see that the cursed thing will come to mind every minute,' he wrote in his *Winter Notes on Summer Impressions* published in 1863.

So where someone is intensely preoccupied by sleep, worrying that they won't get enough and what consequences this will have for the next day, the paradoxical prescription is 'to remain awake for as long as possible'. It sounds crackers, but I discover evidence that this works too.

When it comes to the frequent wakings during the night, O'Regan has several suggestions. One of them is to be performed before you even get into bed and involves a pen and paper. In the 1980s, American psychologist James Pennebaker explored whether the simple act of writing about traumatic events could affect health. He got a bunch of undergrads and split them into different treatments. He had some write about traumatic events they had experienced and others about more mundane matters like a description of their living room, their shoes or a tree. Writing about traumatic experiences like death, divorce or abuse seemed to have a cathartic effect.

'Although I have not talked with anyone about what I wrote, I was finally able to deal with it, work through the pain instead of trying to block it out,' wrote one subject. 'Now it doesn't hurt to think about it.' Pennebaker also found that although the act of writing about these painful experiences caused an immediate elevation in blood pressure and brought on negative moods, it was also associated with fewer visits to the doctor.

Since then, Pennebaker's idea has been put through and

passed more rigorous testing, and a meta-analysis published in 2004 concluded that putting emotional concerns onto paper improved both physical and psychological health. With respect to sleep in particular, there's evidence that it speeds up the onset of slumber.

My favourite method for getting back to sleep though is the 'THE' method, also known as 'articulatory suppression'. In principle, it works a little like counting sheep, getting the brain to perform a task that will distract from thinking about bills, blog posts or book deadlines. But the thing about counting sheep is that they are too distracting, what with their fluffy wool, bleating and bucolic surroundings. Strangely, the act of repeating a word like 'THE', one that is completely devoid of imagery, emotion or colour, seems to shut out the cognitive part of the brain.

Consider a simple bit of maths: add 6 to 13, divide by three and make a note of the remainder. Most people will not struggle to arrive at the answer: 6 remainder 1. But if, each time you exhale, you say 'THE' to yourself, sums like this become almost impossible to execute. I am not sure how well this will work for someone with conventional insomnia, whose mind will be racing in a different way to mine. But whenever I've deployed articulatory suppression to shut down my brain, I have lost consciousness in under a minute.

Cherry-picking from the techniques I'd learned on the CBT-i course to suit my own insomnia – notably sleep restriction and articulatory suppression – I found I was able to realise a depth and continuity of sleep that I can't remember experiencing. After a good night, I would wake to find myself fizzing with energy, as if someone had opened my skull, sprinkled magic dust over my brain and shut the lid. And Lo. Over the course of the day that followed, I could function unbelievably well, with fewer somnolent slumps than normal. In my 20+ years with narcolepsy, the CBT course for insomnia has given me the single most important improvement in my narcoleptic condition.

\* \* \*

I enjoyed the group aspect of the CBT-i course, and there's evidence that this makes it more effective. But if this all sounds a bit much, there's always Sleepio, an online programme delivering a course of CBT-i into the comfort of your own home. This stands out for the very interesting manner in which it came about.

Peter Hames experienced a textbook case of chronic insomnia. 'I had a perfect storm of moving house, changing job, relationship issues and it prompted this period of sleeplessness,' he says. When the stresses eased, however, the sleeplessness stayed. Thinking back to those interminable months, he describes feelings that most insomniacs will be able to relate to. 'I had this immense sense of loneliness, that everyone was sleeping except me. Every night, I'd dread the dawn, looking at the clock every few minutes, but this only served to remind me that I wasn't asleep. I'd be constantly doing the mental arithmetic, working out I had three hours left, two-and-a-half hours left. Every minute that went by made the prospect of the next day ever more daunting.'

Fortunately for Peter, he'd studied experimental psychology at the University of Oxford in the late 1990s, and knew that cognitive behavioural therapy was the leading evidence-based solution for chronic insomnia. He took himself along to his doctor, smugly announced his self-diagnosis and asked to be referred for a course of cognitive behavioural therapy for insomnia – CBT-i. This kind of direct approach does not always go down well with GPs. As far as the doctor was concerned, insomnia wasn't such a big deal. In any case, a talking cure like CBT-i wasn't going to be the solution. 'He refused and gave me sleeping pills.'

It's easy to see why a pharmacological solution is often the first thing that's tried. Drugs are affordable, widely accessible and highly standardised products, which makes it easy to explore their efficacy. A talking therapy like CBT, by contrast, is relatively expensive, requires plenty of expertise on the part of the practitioner and, because it's a face-to-face treatment, it's nigh on impossible to deliver in a standardised package.

Over the course of the twentieth century, drugs for just about every medical condition began to flood the market, creating what some would argue is an addiction-like dependence on pharma. A patient often attends their GP and expects to come away with a prescription. The GP is often happy to oblige. But for some conditions, like insomnia, drugs rarely offer a permanent solution.

Peter went away with his sedatives. They gave him some relief, but it was just one or two nights. The insomnia persisted. He called his sister, a clinical psychologist, and she suggested he read a book by Colin Espie, then director of the Glasgow Sleep Centre in Scotland. If there is one book that gets passed around from one insomniac to the next, *Overcoming Insomnia and Sleep Problems: A Self-Help Guide Using Cognitive Behavioural Techniques* is probably it.

'In six weeks I was totally better,' says Peter. 'Totally cured.' But as he'd worked his way through Espie's book, finding it 'clunky, manual and arduous', he hit upon an idea. In the foreword, Espie explains that there is evidence that CBT for insomnia works, but the means to deliver it is lagging behind.

For Peter, there was a clear solution. All the techniques in the book could be delivered much more effectively through an online package, one that he realised might be able to mimic all of the best qualities of drugs. It would be affordable, accessible, evidence-based and, crucially, it would also be standardised in a way that face-to-face could never achieve.

With his insomnia behind him and excited by this 'epiphany', Peter hopped on a train to Glasgow and proposed a partnership with Espie. They created Sleepio, an online version of Espie's book that would deliver an entire course of CBT-i through the internet. But there was a crucial step in the development of Sleepio that makes it particularly interesting.

Espie knew there was abundant evidence that face-to-face CBT can break a cycle of chronic insomnia. What he didn't know was whether an online course would be effective. He resolved to find out, creating two versions of the course, one that would deliver the evidence-based therapies that he'd written about in his book and the other that would act as a credible placebo. Then he was in a position to conduct a double-blind, randomised, placebo-controlled trial of a psychological treatment. Was CBT-i as delivered by Sleepio's virtual therapist – The Prof – really effective at improving the lives of insomniacs?

In short, yes. The trial demonstrated clearly that Sleepio works. More than that, Espie and his colleagues could put figures on its efficacy. Compared to controls, those receiving Sleepio fell asleep much faster, were awake much less during the night, experienced much better quality of sleep and felt far more alive during the day. Since that first trial, those behind Sleepio have explored the potential for the software to improve other health outcomes, reducing anxiety and depression.

Sleepio is at the forefront of what could be a revolution in the way that healthcare is delivered. With the explosion in the use of smartphones and other mobile technologies over the last decade, it has become possible to collect vast amounts of data about our lives, data that simply didn't exist before. These, in turn, have inspired a field that is being referred to as 'digital health', software that promises to transform these data into applications with real benefits for health.

Some of these are based on nonsense, some on common sense, some on sound science, but there is little way to judge whether these apps really work. We, as consumers, should be wary of this and demand efficacy. Otherwise, we are just handing over cash and personal data in exchange for a bit of software that might help, might not or could even be harmful. Sleepio is one of the first apps whose efficacy has been confirmed through a double-blind, randomised, placebo-controlled trial.

\* \* \*

If CBT-i, either delivered face-to-face or over the internet, can really improve the night-time sleep of people with narcolepsy, then it offers a rather simple explanation for the paradox that narcoleptics often lie awake at night. This is the sleep fuel argument, whereby the daytime naps of narcolepsy reduce the pressure to sleep at night. In my own case, the fact that sleep restriction (as best as I am able to perform it) has improved my sleep at night does indeed indicate that this was at least part of the problem, and almost half those with narcolepsy acknowledge that napping during the day affects sleep at night.

But simple explanations of complex phenomena are usually only part of the picture. In addition to the depletion of sleep fuel, the high frequency of REM in narcolepsy (not to mention the increased prevalence of other disorders like sleep apnea) is very likely to damage the architecture of sleep. We have also seen that the narcoleptic brain may respond to the loss of hypocretins by expanding other neural networks on which these neuropeptides would normally act (like the histamine signalling system, for instance). It's plausible that disturbed sleep might be the price to pay for such accommodation. Or it might turn out that we need to think of the hypocretins rather differently, less as stimulants and more as stabilisers. If their true function were, in fact, to lock the brain into either wakefulness or sleep for long periods then it follows that their absence would result in a continual flip-flopping between these states.

All of these possibilities, and more, could be working together to mess with the narcoleptic's night-time sleep, and once sleep goes out the window there's always the risk that perpetuating factors will wade in, just as they do in many cases of insomnia. In this light, the disturbed nocturnal sleep that often comes with narcolepsy suddenly looks less of a paradox and more of an inevitability.

This helps explain why many of the consequences of narcolepsy, and for that matter every single one of the 60+ possible diagnoses listed in the international classification of sleep disorders, are so similar to what happens when there is little or no sleep at all.

# Mind, body and soul

*'Sleep is that golden chain that ties
health and our bodies together.'*
Thomas Dekker

Sleep deprivation is torture. Literally.

Victorian psychiatrist Lyttleton Forbes Winslow argued that lack of sleep is 'a certain forerunner of insanity!' In a gruesome footnote that appeared in *On Obscure Diseases of the Brain*, he relayed the story of a Chinese merchant who had killed his wife and was sentenced to death by sleep deprivation. In a prison, with three permanent guards on hour shifts to prevent the man from lapsing into sleep, he survived for 19 days, but after a week or so he was already begging for 'the blessed opportunity of being strangled, guillotined, burned to death, drowned, garotted, shot, quartered, blown up with gunpowder, or put to death in any conceivable way their humanity or ferocity could invent'. This, Forbes Winslow noted, 'will give a slight idea of the horrors of death from want of sleep'.

One of the most famous sleep deprivation experiments took place in the 1960s. Randy Gardner, a student at Point Loma High School in San Diego, California, wanted to know what would happen if he went without sleep for a long time. So he roped in two of his classmates and got them to work in shifts, prodding him every time he threatened to nod off. After a few days, he'd started to hallucinate. His speech became slurred. His memory began to fail. He became paranoid. 'I wanted to prove that bad things didn't happen if you went without sleep,' he explained at a press conference held immediately after his ordeal. In fact, after 11 days and 24 minutes without sleep, he'd proved the opposite.

The definitive experimental demonstration of the horrors of sleep deprivation appeared in an infamous paper published by Allan Rechtschaffen and his colleagues in *Science* in 1983. They installed a pair of rats in neighbouring cages. In the bottom of each cage was 3 cm of water, but by standing on a record-player-like disk shared by both cages the rats were able to stay high and dry. The set-up was cleverly arranged so that the animals were subject to the same environment and rotation, made to walk an average of around one mile a day, but got very different amounts of sleep. Experimental rats got almost none, while control rats managed a moderate amount.

With time, the severely sleep deprived rats began to deteriorate, showing at least two of several pathological signs, including ungroomed fur, skin lesions, swollen paws, inability to move, loss of balance and a significant weakening of the EEG signal. Three of eight experimental rats died, one after just five days. When Rechtschaffen and co. carried out necropsies on the deceased they found evidence of further problems, including collapsed lungs, stomach ulcers, internal bleeding, testicular atrophy, severe scrotal damage and swollen bladder. The control rats, by contrast, were in relatively good nick. The conclusion: 'Sleep does serve a vital physiological function.'

This fact is starkly evident in a condition known as fatal familial insomnia (FFI), a horrifying genetic condition that is characterised by a near-total inability to sleep, a pathology that tends to strike in middle age and with little warning. The best-known familial cluster of FFI comes from the Veneto region in northern Italy not far from Venice. In 1836, when he was 45, a man called Giacomo suddenly stopped sleeping properly, and soon he was dead. At least 30 of his descendants have suffered a similar fate, finding themselves in their middle age but completely unable to sleep and dead within months.

In 1984, the family came to the attention of Elio Lugaresi, one of the pioneers of sleep medicine and the founder of the now world-famous Bologna Centre. Lugaresi was able to document the decline of one member of the family and collaborated with Pierluigi Gambetti, a neuropathologist at Case Western Reserve University, to study post-mortem slivers of the patient's brain. This and follow-up work revealed that FFI is a prion disease, with rogue proteins similar to those that cause mad cow disease in cattle and Creutzfeldt-Jakob disease in humans, resulting in degeneration of the thalamus.

The thalamus sits adjacent to the hypothalamus and acts like a gateway between the central, most evolutionarily ancient parts of the brain and the more recently evolved, thinking cortex. Without this portal in place, basic functions go out the window, with profuse sweating and uncontrollable salivation, racing heart rate and soaring blood pressure. Melatonin levels flatline, suggesting a complete breakdown of the circadian rhythm that's so important for the orchestration of sleep.

Although FFI causes such savage sleep deprivation that death comes quickly, it is exceedingly rare, affecting just a few tens of families around the world. But sleep deprivation is very easy to arrange, and the pace of modern society poses many threats to the quantity and quality of sleep. The invention of screen-based technologies has certainly put pressure on sleep. I like TV. I marvel at tablets. Game consoles are so much fun. Mobile phones are virtually indistinguishable from magic. But these devices and the hugely entertaining, near-infinite availability of addictive content threatens sleep, particularly that of children.

In the average UK household, a typical ten-year-old can get his or her fingers on five different screens. If this sounds far-fetched, do the calculation for yourself. In my house of two adults and two children, we have three smart phones, one tablet, two laptops, one desktop and a family TV-cum-gaming screen. With seven screen-based devices, my two sons are well over the national average, with an unbelievable wealth of screen-based entertainment at their fingertips. It has become something of a daily battle to prevent them from overdosing.

It's estimated that one in three infants in America has a TV in their bedroom by the age of one, and toddlers typically spend almost two hours in front of a screen every day. By the age of eight, these figures have swollen, so that almost half of young children in the US have a TV in their room. These stats come from a study published in 2011, so lord knows what they are now. They will only have gone up.

In the UK, older children typically consume an average of over six hours of screen-based media every day. In the US, it's probably more like seven-and-a-half hours. In Canada it's nearly eight. In essence, children in the developed world are spending more than half their waking lives in front of a screen.

In a 2006 study, researchers in Finland surveyed over 300 families with five- and six-year-olds, seeking to understand the children's TV habits and whether this might have any effect on their sleep. What they found was quite alarming. The children were typically sitting in front of the TV for about an hour and a half a day, actively engaged with some cartoon or other. The more a child watched, the less they slept, the more disturbed their sleep and the sleepier they were the next day. In a lot of households, the TV was never switched off, so that even if they weren't watching it, the TV was the soundtrack to many of their other activities. Children exposed to this kind of 'passive' viewing were more likely to struggle getting to sleep.

Sleep is particularly important during adolescence but contemporary teenagers are probably getting less of it than ever before. In a recent study from Norway, the average bedtime for the teenaged subjects was 11.18 p.m., but it was taking them almost another hour to fall asleep, so they were typically getting less than six-and-a-half hours sleep a night. This is around two-and-a-half hours shy of the nine hours recommended for this age group. Many of the youngsters were so shattered during the daytime that a doctor would have diagnosed insomnia.

In another study, published in 2011, researchers from Australia recruited teenage boys to undergo a couple of sleep tests. Before one of these, they were allowed to play 50 minutes on a PlayStation 3, an amount considered normal. On the other occasion, they got to play for 150 minutes.

The boys came to the sleep lab straight from school and in a shared waiting lounge chatted with each other, read and did their homework. They had tea at 6 p.m., got ready for bed and had the wires fixed to their heads. The video game they were given was *Warhammer 40,000: Space Marine*, a rapid action, 'strong violence' game with a rating of 15+. Although lights out happened at the same time on both nights, the prolonged gaming session meant the boys took longer to fall asleep (about 20 minutes or so), they slept less (to the tune of about half an hour) and experienced significantly lower sleep efficiency (falling to less than 85 per cent).

For most people, this kind of sleep deprivation is easily reversed. It just takes a change in behaviour. This can give the impression that losing sleep doesn't matter too much. If it's just one or two nights here or there it probably doesn't. But it's unwise to make a habit of cutting back on sleep. Losing sleep on a regular basis can cause damage in three main ways: through its immediate effects on brain performance; with knock-on consequences for other organs; and frequent psychological consequences too. All sleep disorders can have all these damaging effects, but the one that really stands out is restless legs syndrome (RLS).

\* \* \*

'Imagine you have rack of lamb on your plate, and it's nicely cooked,' says Deborah Henry-Adolph. 'My leg feels as though the flesh is falling away from the bone.' This is how she describes what it's like to live with restless legs syndrome, a sleep-related movement disorder that can quickly lead to a serious case of chronic sleep deprivation.

The rack of lamb is a very strange image, and I'm not sure it helps me to understand the sensation of restless legs. But as I begin to learn about this strange phenomenon and its impact on sleep, it is striking how often people struggle to put the sensation into words. This was a fact noted by the Swedish neurologist Karl-Axel Ekbom, who carried out the first systematic study of the phenomenon in 1944 and put the disorder on the map.

Ekbom had assembled more than 30 sufferers and found them 'frankly puzzled how to depict their ailment'. Many described a 'crawling' sensation in their limbs, others spoke of 'cramps', some referred to 'fidgets'. A few mentioned a sucking, pulling, dragging feeling. One said it felt as though 'something widened and contracted in a slow and irregular rhythm.' Only occasionally did they deploy a simile. 'It feels as though my whole leg were full of worms,' said one. More recently, researchers published a list of descriptive terms that have been used: 'soda bubbling in the veins', 'the gotta moves', 'heebie jeebies' and 'Elvis legs' to pick just a few. It's probably fair to say that one of the defining features of RLS is the difficulty of putting the sensation into words.

What everyone with first-hand experience of RLS has no problem getting across is just how horrific this disorder is. 'It is an unbearable feeling,' said one of Ekbom's patients. 'It is worse than an ordinary disease,' said one woman who had experienced both hepatic disease and a pulmonary embolism. 'I hardly dare to go to bed, it is so horrid,' reported another. 'I get so hysterical, I begin to weep.' Every single person I've ever spoken to about their restless legs says something similar. 'I would not wish RLS on my worst enemy,' is a sentiment I've heard on more than one occasion.

The feeling is obviously intensely uncomfortable, the need for constant movement exhausting, but RLS almost always results in sleep deprivation too, and this surely plays a big part in the dread. The impact on sleep occurs because there is a clear circadian pattern to the restlessness: it intensifies in the late evening and eventually peaks in the early hours of the morning. This strongly suggests that the body's central clock – that molecular circuit in the suprachiasmatic nucleus – is playing a part in this disorder, perhaps even by increasing the levels of hypocretins in the evening.

In the 1990s, researchers quizzed more than 100 people with RLS to get a better idea of its consequences for sleep. Almost all the patients claimed that the constant movements meant it either took them longer to get to sleep, woke more during the night, or both. Follow-up sleep tests confirmed this. To make matters worse, it's estimated that eight out of ten of RLS sufferers also experience another condition known as periodic limb movement disorder (PLMD), regular, uncontrollable muscular contractions, sometimes just a twitching, at other times violent flailing.

'You ever seen a footballer kick a ball? I could play for Manchester United,' jokes Deborah, though it's anything but funny. She has a slipped disk in her lower back and the sudden limb movements can be so severe they cause her agony. 'Sometimes I think my back is going to break. It's as if somebody put a thread or rope around your spine and one person is pulling it to the left and the other to the right.'

These movements jump to a similar circadian rhythm as the restlessness of RLS, increasing in the late evening and peaking in the early hours. Like the excessive REM of narcoleptics or the stop-start breathing of sleep apneics, the restless legs and periodic limb movements tend to interfere with the long stretches of non-REM sleep that would normally occur at this time, bringing a pin to the swelling membrane of a magical restorative balloon. RLS eats into sleep duration. PLMD ensures that any sleep that does occur is so frequently interrupted it's worthless. The upshot, like narcolepsy, sleep apnea, even insomnia, is shattering.

RLS is surprisingly common. Ekbom judged that 'every practicing physician meets it', and he was probably right. In a recent, very large study, it seems as though some 7 per cent of the population has some experience of the phenomenon. Almost half of these people found their restless legs distressing, and most of them – over 300 in the study – had taken their symptoms to their GP.

More than 60 years after Ekbom described this syndrome, one would have thought that most of these sufferers would have come away with a name for their condition and talk of treatment. Strikingly, however, just one in 16 received a diagnosis of RLS. PLMD is common too, thought to affect around one in 25 people, possibly more.

Ekbom died in 1977, when I was just four. But scouring the literature on RLS, it looks like he continued to pen articles on the condition after his death. It is only when I look closer that I realise these posthumous publications are authored by his son, Karl Ekbom Jr, also a neurologist, now in his 80s and retired from his consultant position at the Karolinska University Hospital in Huddinge.

I ask him why RLS should be so common yet so poorly understood, why it's frequently described as 'the most common disorder that you have never heard of'. Part of the reason, he suggests, is that many of those who experience symptoms probably don't have it badly enough to seek medical advice. As with most sleep disorders, there has also been insufficient investment in research into the condition. This means that until recently there was little understanding of what might be causing it, so little idea of how to treat it, and hence not much reason to tell medical students about it during their training. It's probably also true that those who've never experienced RLS can't quite see why a wormy, itchy, crawling sensation should be such a big deal.

\* \* \*

I meet up with Martin Creed in a cafe on Roman Road in the heart of London's East End. It is not a market day, but the street is still bustling, commerce of near-infinite ethnic diversity riding on the same wheeling-dealing wave that has existed here for more than a century. When Martin speaks, his accent and timbre are so extreme that he's almost a parody of cockney, the kind of impossibly husky voice that one might expect to hear on the film set of a Kray brothers' movie. He looks shattered, but then he suffers from both PLMD and RLS and has not slept a wink in three days straight. 'My eyes are burning now through the lack of sleep,' he says.

The first time the restless sensation occurred was in 1982, when Martin was 32. 'I can remember the first night it happened and thinking, "What the hell's this?"' There was a crawling feeling in his feet and ankles that lasted about half an hour. 'I could not sit still,' he says. After that first encounter, the restless legs came and went over the next 30 years, but with each episode the feeling intensified, until around five years ago it seems to have taken root for good, so that now it's every night.

'Once it comes to midday, the dread starts,' he says, hours of anxiety over whether the restless legs are going to pay a visit that evening. 'It starts around 9 o'clock when it's time to go to bed.' The only thing that stops the sensation is to move. 'You have to do it.' During a typical night with restless legs, Martin is up and down like a yo-yo. Sitting is no good. 'You still got it.' Making a cup of tea helps. Walking is better. 'What I tend to do is go in the garden and walk up and down.'

If he does manage to get to sleep, the PLMD will often kick in, quite literally. 'When I jump off the bed, I wake for a second,' he says. After a couple of successive nights with RLS and PLMD, the exhaustion is overwhelming.

The underlying cause of both RLS and PLMD remains somewhat mysterious. In the case of RLS, Ekbom spotted that it appeared to be especially common in anemia and in women during pregnancy, both of which are characterised by low levels of iron. There is fairly wide consensus that an abnormality of iron metabolism plays a role in many cases of RLS. There is evidence too that there could be a problem with the dopamine pathway in the brain, the same neurons that degenerate in Parkinson's Disease, resulting in muscular twitches and shakes. Drugs that boost dopamine activity tend to improve RLS, while drugs that block the same system make it worse. Interestingly, these two observations – iron deficiency and dopamine signalling – may be two sides of the same coin, because iron is needed for the production of dopamine.

As we discovered in chapter five, the hypocretin neurons in the brain have a major modulating effect on the dopamine system. This may explain why around one in five people with narcolepsy meets the diagnostic criteria for RLS and almost one in two has limb movements in their sleep (though these tend not to come at regular intervals, as they do in regular PLMD). I am fortunate I do not suffer from RLS (yet), but I have, on many occasions, done battle with snakes, dinosaurs and lions in my sleep. I twitch and whimper, sometimes even shout and kick out.

Whatever is underlying these disturbing movement disorders, they can have life-changing, sometimes life-ending consequences.

\* \* \*

The most immediate, most obvious impact of sleep deprivation is on brain performance. 'If a man is denied sleep so that the drain upon nerve cells continues beyond a certain point, he will, of course, be thrown into a condition of fatigue, when intellect and emotions must suffer,' wrote Michael O'Shea in *Aspects of Mental Economy* in 1900.

In order to gauge the extent of this damage, psychologists will often deploy the psychomotor vigilance task. This requires a subject to strike a button every time they experience a stimulus, like the flash of light on a screen or a beep from a loudspeaker. The stimuli occur over and over but at random intervals over the course of a ten-minute test. Sleep deprivation seriously messes with the ability to carry out this task, slowing the reaction time and increasing the number of lapses. When someone doesn't even respond to one of the flashes or beeps, it's probably because they were so tired they lapsed into one of those microsleeps that are common for many people with narcolepsy.

Sleep deprivation and the frequent failures of consciousness that result are like taking a wrecking ball to memory. 'One of the cruellest things about having narcolepsy since very early childhood is that I have few memories and the ones I have are actually "false memories",' says Lily Clarke. When she turned 40, her best friend gave her an album of photos from their youth. 'I burst out crying and everyone thought I was being sentimental but I was inconsolable because I didn't recognise any of the events.'

Most people with narcolepsy report something similar. 'I have holes in my memory,' says Pen Pearson. 'It's like going to open a drawer for something and when I open it the drawer is empty.' For Trish Wood, her short-term memory is definitely affected. 'Dates and names are a particular problem,' she says. 'I am constantly late for work because I have lost or forgotten things like keys, medication, handbag, purse and lunch. I lay everything out the night before but can still manage to lose my car keys in the short distance between my kitchen table and the front door.'

Although I am fortunate that I do not suffer from narcolepsy anywhere like as badly as this, I have still become acutely aware that my short-term memory isn't as good as it should be. I have a pretty good recollection of my youth; the memories that were laid down in my pre-narcoleptic days are virtually untouched and readily accessible. I do remember lots of things from my life with narcolepsy but I have a nasty feeling that they are not nearly as sharp as they should be. In fact, it's almost as if I am living in a spotlight focused on the present, my immediate past and future out there in the darkness. It takes a lot of effort to recall what I did last weekend or what I'll be doing next weekend. In conversation, I will often struggle to find the right word, halting mid-sentence for too long. Occasionally, I will have absolutely no recollection of a conversation, an experience that is as alarming as it is embarrassing, and one I did not imagine I would have until I was considerably older than I am now. In my early 40s, I began to contemplate that these memory deficits might be a sign of early-onset dementia. The reality is they are much more likely to be down to deprivation of proper sleep.

Sleepiness affects working memory, the memory needed to process in real time, like when holding several numbers in your head. It is also important for the formation of more long-lasting memories. In simple terms, the brain transforms an experience into a memory in two discrete steps, encoding an event as a temporary memory and then consolidating it into long-term storage. In order to do this well, it is important to have good sleep both before and after an experience.

I go to a lunch party. If I have not slept well the night before, my brain will not be in optimal shape to encode the names and faces of the guests into my short-term memory. If things are really bad, I might slip into microsleeps too, failing to register meeting one or two people at all. Even if I have slept well the night before and bank plenty of information into my short-term memory, I can only store it away for the long term if the night after the event is also good.

The lapses in cognitive function and the microsleeps that result from poor sleep can obviously cause accidents. 'You've got a cup of tea and it falls out of your hands and burns your legs,' says Martin, recounting an everyday hazard of being permanently shattered. He's honed a technique for drinking his tea safely when he's tired, leaning on the kitchen table with his elbows, his hands around the cup. Then, if he falls asleep before he's finished the mug, he won't drop it.

When sleep-disordered or sleep-deprived people find themselves in positions of responsibility, the consequences can extend to others. Surgeons, for example, are likely to struggle more with an operation when they are sleep deprived. In one study, researchers used a virtual reality set-up to test the dexterity of trainee surgeons in different stages of sleep deprivation. Those who'd been awake all night made 20 per cent more errors and took 15 per cent longer to complete the screen-based operation than those who'd had a full night's sleep.

The meltdown at the Three Mile Island nuclear plant in Pennsylvania in 1979 and the Chernobyl disaster in the Soviet Union in 1986 are always cited as accidents where sleep deprivation resulted in human error, though the evidence is circumstantial. In the case of the Space Shuttle Challenger accident in 1986, however, the presidential commission ruled that sleep loss and shift-working played a part.

As horrific as these large-scale tragedies might be, the devastation they cause is nothing compared to the far more mundane problem of drifting off at the wheel. It's been estimated that staying awake for 24 hours results in an impaired cognitive state roughly equivalent to having a blood alcohol level of 0.1 per cent. In most countries, this exceeds the legal limit for driving.

Though it's obvious that sleepy driving should increase the risk of an accident, it is far from easy to figure out just how common such events might actually be. Jim Horne, who until his retirement was head of the Loughborough Sleep Research Centre at Loughborough University in the UK, has spent more time than most thinking about the dangers of sleeping at the wheel. Some 20 years ago, he was approached by the Devon and Cornwall constabulary, then by other police forces around the United Kingdom, each of them giving him access to their records. His mission was to report back on just how common sleep-related accidents might be and whether there were any interesting patterns lurking in the data.

Working with his colleague Louise Reyner, Horne devised a way of homing in on crashes where sleep was the probable cause. As the dataset only involves those accidents reported to police, it can only give us a partial picture of the severity of this problem. But it does reveal some key, take-home messages about sleep-related road accidents.

Most strikingly perhaps is the straight-forward observation that these accidents are more likely to be fatal than other kinds of prang. 'One of the cardinal signs of these collisions is there's no sign of braking beforehand,' says Horne. 'We think there are more fatal accidents associated with falling asleep at the wheel than with alcohol.'

Over the course of a 24-hour period there are two clear spikes to the incidence of accidents where sleep was involved, one in the early morning and a second mid-afternoon. Horne puts the early-morning spike down to 'a triple whammy'. If you are on the road before 6 a.m., it's possible, even likely, that you haven't had enough sleep. This is also the time of day when the body is at its circadian nadir, and all-round alertness is likely to be compromised. To make matters worse, the roads are dark and usually empty, so don't have much to offer in the way of stimulation. 'That's why half the collisions on our motorways between 2 and 6 o'clock in the morning are sleep related,' says Horne.

The second peak in sleep-related accidents is down to the circadian slump in the middle of the afternoon, but the roads are busy so it's not as obvious as the first. 'There's a lot of stimulation going on and stimulation helps offset sleepiness.'

In terms of who is behind the wheel when these accidents occur, Horne identifies three groups that are at particular risk: men under the age of 30. 'They drive faster, cut corners, show off a bit, seem to think they are invulnerable and are more likely to be driving in the early hours of the morning,' he says; shift workers, particularly after the first night shift before their body clocks have adjusted to being awake at night; and those with undiagnosed obstructive sleep apnea.

There's good evidence, for instance, that treatment with CPAP significantly reduces motor accidents. In a study that took place in Ontario, Canada, researchers found that before treatment, the rate of accidents for the obstructive sleep apnea patients was three times that of the general population. After treatment, the accident rate dropped to normal levels.

Motorways are particularly dangerous. The unwavering, high-rev thrum of an engine and the sheer monotony of pummelling along in a straight line making sleep that much more likely. Owing to the fact that the opportunities to get off a motorway are limited, there will often be a period where a sleepy driver will have to push on for many miles in a sub-optimal state, rolling down the windows, switching the air conditioning to freeze, shouting or singing along to a maxed-up radio.

Horne has looked into whether these countermeasures

work, specifically cold air and the radio. 'Cold air in your face can have an effect for a short while, maybe 15 minutes,' says Horne. 'It depends how sleepy you are.' The radio, however, is 'basically useless'.

'As soon as you start doing things to keep yourself awake you know you are a danger to yourself and other people,' he says. 'You shouldn't start thinking I'll turn up the air conditioning and I can carry on driving for an hour or so.' It's time to get off the road at the next stop.

Horne's work led the UK government to introduce signs on motorways around the country, stating the obvious but easily ignored mantra: 'Tiredness kills. Take a break'. He has also carried out research on the most effective way to take that break, giving a group of graduate students either a strong coffee or a caffeine-free control and a nap of up to 15 minutes before testing their driving performance in a virtual test.

Buy a coffee, one containing a hefty amount of caffeine. 'Drink the coffee and go straight back to your vehicle,' he says. 'As the caffeine takes around 20 minutes to kick in, here's that window of opportunity to get your head down for a quick zizz. Even if it's dozing it can be quite refreshing.' The caffeine buzz and a short doze combine to provide a powerful pick-me-up.

Most people with a sleep disorder understand the dangers that sleeping at the wheel could pose to themselves, their passengers and other road users, and take it seriously. Some will not drive at all, but many – with the proper medication and management – can continue to do so safely. 'I have never come across a case of someone with narcolepsy, diagnosed and treated, ever having a serious collision on the roads as a result of falling asleep at the wheel,' says Horne. In fact, given the self-imposed rules that many people with sleep disorders have to safeguard against sleep while driving, it's even possible that they are safer than many other road users, he says.

\* \* \*

In the longer term, sleep deprivation can have serious consequences for the body. For over 20 years Francesco Cappuccio has worked as a consultant in cardiovascular medicine, flipping between academic research and clinical practice at the University of Warwick. Around 15 years ago, he began to ask questions about the long-term health consequences of insufficient sleep. He was specifically interested in those who huddle beneath the lower lip of the sleep duration bell curve, the one in eight people who get less than six hours sleep a night. Could routine short sleep be contributing to the obesity epidemic unfolding across the developed world? What about cardiovascular disease? Type 2 diabetes? Are those with short sleep more likely to experience an early death?

Cappuccio and his colleagues pooled data from 30 research papers linking sleep duration with obesity. 'All the studies published at the time indicated a very significant association between the proportion of people that are obese and sleep duration,' he says. But an association between these two variables falls well short of demonstrating a causal link. It is relatively easy to see how obesity, by increasing the fat laid down in the throat, could cause sleep apnea. Is it possible that it could also work the other way round, with short sleep somehow causing obesity? Cappuccio resolved to data mine his way to a conclusion.

The longitudinal study is of crucial importance, one in which data are repeatedly collected from the same individuals over the course of many years. This can help address the question of which came first, the short sleep or the obesity. A longitudinal study of young children in New Zealand was the first to indicate that it's the short sleep that kicks things off. Cappuccio's own data, as yet unpublished, shows much the same. 'We are convinced that the exposure to short sleep precedes obesity,' he says.

In the context of sleep apnea and narcolepsy, we have already seen how disrupting sleep can have unhealthy consequences for metabolism. Getting too little sleep does much the same, and possibly more. There is now compelling evidence that chronic sleep deprivation doesn't just lead to obesity, it is also associated with a long list of other health complications, including an increased risk of type 2 diabetes, high blood pressure, structural damage to the blood vessels, stroke and coronary heart disease, to name just a few. People with habitually short sleep are also more likely to die early.

* * *

On top of the immediate effects on brain performance and longer-term damage to the body, chronic sleep deprivation can take a serious psychological toll. The bullying and mockery heaped on the sleepy is by no means confined to children, but this is when it is most obvious. It's a simple fact that people are quick to make moral judgements about someone who is sleeping when others aren't.

In 2012, aged 80, Ronald Embleton sent me a series of handwritten letters – over 50 A4 pages in all – describing his early life in extraordinary detail and the onset of his narcolepsy. Looking at his beautiful handwriting and reading some of his colourful memories, I see the work of a man who knew he was dying but one excited at being able to share his experience of his sleep disorder. I would have liked to have met Ron very much, but sadly he died before I got the chance. I am sharing his story because that is what he wanted, but also because it illustrates perfectly the psychological toll of a slow, bewildering descent into a life of sleep in one too young to know what was happening.

In 1941, when Ron was nine, he took the equivalent of today's 11-plus exams. He was evidently bright, coming in the top six in the borough of South Shields in the northeast of England. 'I could have picked a place in any high school around,' he said. 'Offers came from Sunderland, Newcastle and, of course, from South Shields.' As Ron was still young and the war was still raging, there was no way his parents were going to let him travel as far afield as Sunderland or Newcastle, so he settled instead for a small church school closer to home. But it was then, probably when Ron was around ten, that he began to notice feeling unusually tired during the day. The family had an allotment, where they grew potatoes and other vegetables to supplement their war rations. 'We were expected to help in the preparation work once or twice a week,' digging over the ground and so on. 'This would virtually kill me,' he remembered. His older brother Bob was always up for running home but Ron was surprised to find he couldn't. 'I would have crawled if allowed to.' The physical exhaustion that Ron describes could easily have been down to undiagnosed myalgic encephalomyelitis or chronic fatigue syndrome, which is one of many co-morbidities suffered by people with narcolepsy.

Ron began to struggle to read, 'not because of my eyesight or ability, but tiredness. The lines wouldn't stay straight – the print bounced and blurred, and suddenly there would be a bang and the book fell on the floor when my hands lost their grip.'

Bewildered and uncertain what was going on, Ron turned to his parents, but they simply put his exhaustion down to a combination of adolescence, overexertion and late nights. It was a phase that would pass, or so he was told. With no one to turn to, 'I would break down and drown myself in tears,' he said.

At the age of just 14, Ron left school to take up a position as a colliery engineer in the dark, dingy and dirty drawing offices at the Harton Coal Company. Compared to most of his contemporaries, who went straight into the colliery workshops, Ron had landed 'the soft job', one in which there was no excuse for being sleepy. So when narcolepsy swept over him on most afternoons it caused him great anxiety.

Worse was to come. At the age of 18, Ron was transferred to Whitburn Colliery to get practical experience in mining engineering, a period he described as 'three years of hell'. The workshops were filthy, the floor and workbenches uneven and impregnated with grease and coal dust. Most of the machinery was powered by steam. Old equipment lay abandoned in every corner. There were no overhead cranes, so every part, every product, every piece of machinery had to be manhandled into and out of the shop. To make matters worse, the other workers resented the arrival of the smart office boy and were quick to pick up on his sleepiness. Work began at 7 a.m. sharp, with a breakfast break at 9 a.m. and on one occasion, Ron fell asleep eating his food, 'something the others never allowed me to forget'.

'I felt degraded and embarrassed to let anyone see me sleep,' remembered Ron, so he began to seek out quiet spots he could sneak off to for a rejuvenating rest. In winter, the steam houses were 'wonderful places'. In summer, he'd slip into a corn field at the back of the workshop, 'make a bed like a cat or dog'.

There was also a scrap area that he liked, where he could hide himself in the long grass. 'The weather or temperature didn't stop me sleeping, only made me a bit more uncomfortable when coming to wet or freezing cold.'

Life in the workshop was hard enough, but when Ron was called out to conduct a repair at the coalface, it was considerably worse. When a bit of machinery went out of action the mine managers pushed the engineers to resolve the problem in as short a time as possible. This meant 12-hour shifts and a long and somewhat unorthodox journey to work. A caged lift would drop down the mineshaft, often accompanied by streams of rainwater and cold air from the surface. At the bottom, the men would climb into empty coal tubs and sit, legs crossed and hunched, for a trundling rail ride far out under the North Sea. 'It rocked me to sleep almost every time,' said Ron. At the coalface – after a journey of up to an hour and a half – the air was stale and smelly and the surfaces damp with seawater seeping in from above. In these closed and intimate conditions, Ron could not hide his sleep from his fellow engineers and miners. There was nothing for it but to nap in the cage, in the coal tubs, in a small alcove in the mine, in an engine house.

The cumulative psychological impact of these years of strain is evident in how Ron responded to the National Coal Board's offer of a scholarship to study engineering at university. He turned it down. 'I could not expose myself completely in a world among strangers,' he said. 'I couldn't possibly face being away from the protection home gave me at the time.' The misery led him to think of how best to kill himself. 'I was definitely feeling like ending the fight altogether.'

It might well have been talk of suicide that resulted in Ron's GP finally taking his symptoms seriously, referring him to see neurologist Frederick Nattrass at a hospital in Newcastle.

After two weeks of being poked and prodded, 'a guinea pig for students and trainee doctors to practice on', Ron received a diagnosis of narcolepsy. It was 1953, he was 21 and had been living with the condition – undiagnosed – for more than ten years.

Decades have come and gone, sleep apnea has put sleep science on the map, we understand a lot about the central importance of the circadian rhythm and we have even discovered the neurological basis of narcolepsy, but people are still quick to judge, then bully, the oversleepy.

Anders Lauszus is the youngest of three children living with his family in Kolding, a small seaport town in the south of Denmark. The first indication that something was wrong came when Anders was just nine and about a week after receiving Pandemrix in 2010. The vaccination brought on a fever, but though this didn't last long, he could feel that something had changed. He was struggling to stay awake during lessons. 'I remember sitting there, at the back of the class, and thinking the lesson was really boring,' he says. This might sound like a fairly typical experience for a young boy, except that in the boredom, his brain began to shut down. 'I was looking down at the book we were reading, following the text, when all of a sudden I became tired. My eyes began to feel really heavy. I looked around confused. It was as if I was in a dream.'

This was not a one-off, and it wasn't long before the teachers began to notice. 'I would be told to cut it out and start paying attention to the lesson.' One of the teachers imagined that Anders must be staying up too late at night when, in fact, his immune system had just razed the hypocretin neurons from his brain. 'Every night when I was doing homework, my parents were constantly waking me up, asking why I was falling asleep, why I was so tired,' says Anders. 'I didn't know. I couldn't think straight. I simply couldn't.' He found himself unable to fulfil the 20 minutes' reading that his school expected every day. 'I would just stare at a page in my book,' he remembers, turning pages for the allotted time 'as if I were brain-dead'.

The first time his friends witnessed Anders in the throes of a cataplectic attack they didn't think too much of it. But as the fits strengthened, he repeatedly found himself at the centre of unwanted attention. There were boys in his class who began to pick on him because he was standing out. The girls showed more concern but it didn't stop them from laughing. Transitioning from primary to secondary school, hoping that it would give him the chance to make a fresh start away from most of his primary peers, it wasn't long before he was being picked on again. There was one child in particular – Christian – whose bullying led Anders to retaliate with a headbutt. The incident saw him changing schools yet again, this time to one that was 'more peaceful', where staff and children had been made aware of narcolepsy and the challenges it poses to learning.

On a continent like Africa, where there is still less understanding of the importance of sleep and the impact of sleep disorders, the bullying can take an unusual form. Jane Wachera is almost exactly the same age as Anders, but in 2010 was living in Nakuru, the fourth largest city in Kenya. When she fell asleep in class, the teachers at her school came down hard on what they perceived to be laziness. On more than one occasion, Jane awoke at her school desk to find that her fellow classmates had drawn a sleepy cartoon of her on the blackboard, embellishing it with unkind comments.

Her mother Anne responded by transferring Jane to a school in Kabazi, a small town some 30 miles north of Nakuru where her own mother (Jane's grandmother) was living. Anne assumed that in the close-knit village community the bullying wouldn't be so intense, but it soon transpired that if anything it was worse. 'Everyone knows who is who,' says Anne. 'If someone has a problem, they will definitely pick on you.'

Jane slept a lot at school and was subject to the usual abuse. What was almost more of a problem, however, was that she performed consistently well, frequently coming top of the class. For the teachers, there could only be one explanation: Jane was bewitched, a contemporary echo of what Massimo Zenti experienced at his Catholic school in Italy in the 1980s. Word soon got around. 'The village people would tell her she was devil worshipping,' says Anne. When teachers woke Jane and, hoping to embarrass her in front of her classmates, demanded to know the content of the lesson, they were confounded when she was able to tell them. 'They didn't know that when she sleeps she can hear. It was like magic.'

The same thing happened at church. Jane would sleep, but was still able to relay to others what the pastor had been talking about. Her friends began to fear being touched by her. Even at family gatherings, everyone avoided her. 'She became really stigmatised to the extent that she stopped wanting to go to school, to associate with anybody,' says Anne.

Back from school one lunchtime, Jane sat down to write a message that she intended to be her last. In it, she wrote how she could not understand why she slept while others were awake, why nobody loved her and why there was so much hatred towards her. Ironically, it may have been Jane's narcolepsy that stopped her from taking her life. When her grandmother came home to cook Jane lunch, she found the child asleep at the kitchen table, pen still in hand and dribbling over her own half-finished suicide note.

Anne rallied behind her daughter, setting up Narcolepsy

Awareness Kenya and the wider Narcolepsy Africa Foundation and campaigning for greater awareness of narcolepsy across the African continent. 'I could have lost her because of ignorance or maybe not taking things seriously,' she says. There are few data on the prevalence of narcolepsy or sleep disorders in Africa, but from Anne's work it's clear that there are a lot of people in need of help but few places to get it. 'I tell them it's real, it's not down to their caste, it's not witchcraft. It's just a neurological brain disorder,' she says. Jane's ambition is to become a neurologist and set up a sleep clinic in Kenya.

These stories touch on what might be the most serious result of sleep deprivation: deterioration of mental health. In the past, in the days when people weren't paying much attention to sleep, it was clear that depression could interfere with sleep. Nowadays, it's evident that the relationship between sleep and depression goes both ways. In many cases it's the poor sleep that comes first.

For Martin Creed, the sleep deprivation caused by his RLS and PLMD will often bring on an episode of bipolar disorder. 'I started to neglect myself, not eating and gradually losing weight,' he remembers of a sleep-deprived, depressive phase shortly after his marriage. 'My clothes looked like they'd been handed down from a big brother.' Transitioning to a manic phase he went on a spending spree for new clothes. He came home with around ten expensive new shirts. 'I looked better and it made me feel better but the feeling was short-lived.' As he continued to lose weight, reaching a skeletal 58 kilos at the low point, the new shirts no longer fitted so he bought still more. 'I was trying to fix myself by buying clothing.'

After three consecutive days without sleep, Martin was in a very bad place and his wife had never seen him like it. 'She was frightened of coming home and finding me dead.' So Martin took matters into his own hands, climbed into the car and set off for Beachy Head, the highest promontory on the white cliffs of Dover and an infamous suicide spot. 'I was going to drive straight off it,' he says. 'I'd had enough. I didn't want anyone to find me.' Halfway there he got a call from his wife, who managed to get him to see through the combination of sleep deprivation and bipolar that was clouding his judgment. 'I had a moment of clarity, a moment of sanity,' he says.

Many people with narcolepsy will recognise the consequences of poor sleep on their psychological makeup. For those with narcolepsy *and* cataplexy, the cataplexy often has another impact that is less likely to endanger life but is still profound. In my own life, this role for cataplexy is most evident in my relationship with sport. It wasn't long after I'd experienced cataplexy in response to Rob Andrew's drop goal that I began to notice my muscles going in other sports, particularly ball games like cricket, football, tennis and volleyball, where the anticipation, excitement and elation of competition would frequently get the better of me. As I pictured the perfect shot in my mind's eye, I would feel a momentary slackening of muscle tone in the milliseconds prior to making contact, just when I needed to be at my most resolute. I found it almost impossible to strike a ball with anything like the power I'd been used to.

In the first 20 years of my life, I loved being part of a team, the camaraderie of rugby, hockey, cricket or the intense competition of squash and tennis. I could do anything. Post-narcolepsy, I did not notice it happening, but I began to withdraw from these social sports in favour of solitary ones. On my own, without an audience with which to share my emotions, I am unlikely to suffer from cataplexy. In the grand scheme of things, this is not too much to bear, but it is a powerful illustration of what many people with cataplexy know only too well: the best way to avoid an attack is simply to steer well clear of strong emotions.

Josh Hadfield – just ten when we first met – recognised precisely this transition in himself. I was asking him about cataplexy and whether making a joke was likely to bring on an attack.

'I don't actually make jokes any more,' he replied, completely matter-of-fact.

'You do sometimes,' his mother Caroline countered.

'Yeah,' he conceded. 'But very rarely.'

Caroline continues to try to find ways for Josh to be playful, humorous and cheeky, but she knows that he has changed. 'He will only joke and laugh if he feels very, very safe and very, very comfortable around people,' she says. 'He will control everything. I hate it. I hate the fact that he can't just laugh out loud at things, that he can't joke without having to think about it or just go off and do something that might be fun.'

Anders Lauszus was also aware of being changed, finding himself increasingly uncomfortable in social situations, 'like part of my brain was telling me that everything connected with communicating and socialising with other people was bad'. By the time he reached sixth grade he'd gone from being outgoing and gregarious to something of a recluse. He learned to avoid situations of emotional excess. 'If I sound like I'm laughing, I'm not really,' he says. 'I'm acting.'

Dee-Dee's cataplexy was so extreme that he quickly became fearful of leaving the safety of his bedroom. 'I learned to control my emotions. I became a fake person,' he says. 'I wasn't me any more.' Still living with his parents, his world was not much bigger than the bed he slept in. If friends were round and cracking jokes, he'd put his 'shields up'. One of the most effective methods to divorce himself from the swirling humour that threatened to floor him was to start counting out long numbers in his head. 'As my cataplexy got worse, it stole more and more of my life, more of my real personality.'

Finally, there is one question about cataplexy that I get asked a lot. Does it happen during sex? In this respect I am fortunate. It does not happen. But for some with narcolepsy and cataplexy, this is a reality. In fact, our old friend the barrel-maker Monsieur 'G' probably suffered from so-called 'orgasmolepsy', his son 'conceived in a moment when the illness came over him'. In her book *Wide Awake and Dreaming*, Julie Flygare revealed that 'being sexually aroused left me limp as a ragdoll', and has written more about this since. 'With orgasms, my head would start falling back like I had whiplash.'

Dee-Dee is frank about the impact that cataplexy has had on his personal life. He knows what it's like to fall in love, but it has only ever happened once. He was 21 at the time. 'I was doing OK at first,' he says, recalling how he was able to hide it. But over the months that followed, the emotional intensity of being in and making love led to more frequent attacks. He was typically having over 50 attacks a day, just through the elation of it all.

On one occasion, while Dee-Dee and his girlfriend were having sex, he experienced a full-on attack, sinking onto her like a dead weight. 'She was totally trapped by my body and found it hard to breathe,' he says. This was not a fleeting fit either, but lasted for several minutes.

After this episode, Dee-Dee went to see the doctor and his girlfriend was supportive at first, but without answers his only solution was to put up barriers. In the end, at some point that Dee-Dee finds it impossible to define, cataplexy robbed him of his capacity to love. 'I've had to let go of the dream that I will ever be with a woman again,' he says.

Superficially then, cataplexy might look like a bundle of laughs. But for many people, it has a cumulative and chilling effect upon life. The world before cataplexy is populated by an infinite winking starry universe of possibility. The world after cataplexy is finite. Sometimes, the restrictions it imposes are not too great and there are still significant areas to explore. For some people, however, cataplexy can extinguish so many stars that there is more darkness than light.

# Good sleep

*'I want my hypocretins back.'*
Henry Nicholls

Ever since the late 1930s, when the pharmaceutical company Smith, Klein and French produced an amphetamine under the brand name Dexedrine, people with narcolepsy have been prescribed simulants in an effort to counter the kind of excessive daytime sleepiness that has plagued my life since my early 20s.

Tony Broad was one of the first people with narcolepsy to benefit from a prescription of amphetamines. Born in 1922, he began to show signs of sleepiness in the late 1930s while studying at Acton Technical College to the west of London. In 2012, I visited Tony, aged 90, at his home near Oxford. 'I remember being sleepy in class,' he told me of his years at college, but because he was a gifted engineer he landed a job in the research department of the music giant EMI. When war broke out in 1939, he and his colleagues began work on transforming a radar signal into a blipping dot on the screen. 'This was regarded as essential to the war effort.' The sleepiness, or 'the dreaded lurgy', as he'd begun to call it in the early 1940s, still hounded him. But within a few years of onset, he had the fortune to be seen by one of the only sleep specialists in the UK and came away with a diagnosis of narcolepsy and a prescription of amphetamine.

It is common for people with narcolepsy to discuss what medication they are on, because these are many and varied, are sometimes effective and often not at all. After years of trial and error, people figure out the dose that works best for them and precisely when to take each tablet to get the maximum benefit. By the time I was diagnosed with narcolepsy in 1995, amphetamines were no longer being prescribed as the stimulant of choice, but Tony was keen that I should try them. 'One tablet a day won't hurt you to do it and it definitely will give you a lift,' he said. 'Can you get hold of it?'

It could be tricky, I explained. 'I can give you a couple now to try to see if you feel better on it. I wouldn't say it will render you bright and new. It won't. But it will make life a lot better and you might live into your 90s.' Tony passed away in 2015, aged 93. He'd been taking these drugs every day for more than 70 years. I still have the blister pack of amphetamines that he palmed me that day in 2012, but it lies unopened in a drawer. Instead, I have been prescribed other stimulants, eventually finding a regime of methylphenidate and modafinil that produces the best results for me.

In addition to stimulants, people with narcolepsy are often prescribed another drug or drugs to control the cataplexy, the sleep paralysis and the hypnagogic hallucinations. In some cases, a small dose of a simple antidepressant like clomipramine can pull this off. For decades, nobody had a proper explanation why such a simple treatment was so effective, but it is now becoming clearer. A few years ago, Takeshi Sakurai, a neuroscientist at the International Institute for Integrative Sleep Medicine at the University of Tsukuba in Japan, demonstrated that a region of the brain that deploys serotonin (called the dorsal raphe nucleus) needs to be working well to prevent a surge of emotions activating the muscle paralysis pathway at the base of the brain. Without hypocretins, the dorsal raphe operates below par, hence the cataplexy. Antidepressants, which boost the serotoninergic signalling, go some way towards compensating, making cataplexy (and probably the sleep paralysis and hallucinations too) less likely to occur.

In recent years, a different and somewhat paradoxical pharmaceutical approach to narcolepsy has emerged. Sodium oxybate (branded as Xyrem) is a drug with a similar structure and action to the powerful sedative gamma-hydroxybutyric acid, or GHB, a chemical that has been used by rapists to spike the drinks of their victims. For people with narcolepsy, however, sodium oxybate not only promises uninterrupted sleep but it offers improvements to both the excessive daytime sleepiness and cataplexy too, a fact that adds weight to the idea that narcolepsy is about too little sleep at night as much as it is about too much during the day.

But all these drugs – the stimulants, the antidepressants and even the sedative sodium oxybate – were invented before anyone knew of the existence of the hypocretins. Evidently, therefore, they were not designed to repair the very specific pathology that appears to explain most cases of narcolepsy, but are used simply because they help manage some of the symptoms. So even with a diagnosis and medication, living with narcolepsy is still a considerable challenge.

\* \* \*

The Epworth Sleepiness Scale is commonly used to assess the extent of the problem. It's a sleep-based quiz to estimate the chances of dozing in eight different settings: while sitting and reading; watching TV; sitting inactive in a public place, like in a theatre or a meeting; as a passenger in a car for an hour;

lying down to rest in the afternoon; sitting and chatting to someone; sitting quietly after a non-alcoholic lunch; and in a car while stopped for a few minutes in traffic.

In each of these situations, you score yourself according to the following scale:

0 = would never doze

1 = slight chance of dozing

2 = moderate chance of dozing

3 = high chance of dozing

Totting up the points for each of the eight questions will give a single ESS score that lies somewhere between 0 (where there's no chance of dozing in any of the activities) and 24 (high chance in every eventuality). A total score of 10 and under is considered normal; 11–12 suggests 'mild' excessive daytime sleepiness; 13–15 is 'moderate'; and 16 and over is 'severe' excessive daytime sleepiness.

Even with my cocktail of stimulants I score around 13, so still suffer 'moderate daytime sleepiness'. I have identified some vulnerabilities. I always get sleepy in the afternoon at around 4 p.m., presumably before my circadian clock has got round to delivering its alerting signal to counteract the relentless rise in sleep pressure. Eating too much for lunch is rarely a good idea. Stepping onto any form of public transport will always result in a rapid voyage to the land of nod. This particular slumber is of a light, vigilant kind, and I usually wake up at every stop to take stock of where I am. It is not uncommon for me to overshoot, however, sometimes winding up at the bus terminal or the end of the line. If I resist the urge to sit, I have a greater chance of staying awake, but it's by no means guaranteed, and a standing nap is often required.

Where sleeping during the daytime is just not an option, most people with narcolepsy will try other tricks: star jumping, arm pinching, face slapping, head dousing, gum chewing. The cold can be particularly helpful. At Narcolepsy UK's annual meeting a few years ago, I encountered one delegate in the gentleman's toilet particularly keen not to miss a moment of the fun, his sleeves rolled up, running his hands under the cold tap and methodically dabbing his forearms with a wet paper towel. At any other event this might have looked decidedly odd. At a narcolepsy knees-up, this is par for the course. Dousing the skin accentuates the temperature difference between the warm inside and the cool exterior. The greater this gradient, the higher the levels of noradrenaline circulating in the blood, a hormone that acts on several regions of the brain to enhance arousal and vigilance. But countermeasures like these are, at best, (exceedingly) temporary measures.

Popping an extra pill can sometimes give more respite, but this requires some planning, because the fastest-acting stimulants take around an hour to kick in. There are other drawbacks of taking extra meds. Often, for some inexplicable reason, the bonus dose simply has no effect. More medication can mean more side effects too. These can be immensely variable, including (but by no means limited to) nausea, headaches, anxiety, moodiness and a very live temper. Sometimes, the side effects are so heinous that it's simply not worth double dosing. In the case of sodium oxybate, for example, taking an extra dose of such a strong sedative could be very dangerous, even lethal.

The bottom line is that there is nothing we have yet devised that comes remotely close to repairing the neurodegeneration that underlies narcolepsy. I want my hypocretins back.

\* \* \*

On the internet there are any number of companies with

synthetic hypocretins for sale. The going rate for 1 mg of white, powdered hypocretin is around $200. Such is my craving that I am tempted, but I know I'd be wasting my money. If I were to sip from the vial of lab-made hypocretin, the enzymes in my gut would pull it apart in no time, snapping the protein necklace and scattering its amino acid beads in all directions. Even if I were to inject the hypocretin into a vein it still would not reach my hypothalamus.

The blood–brain barrier exerts very strict control over what passes from the blood to the brain, rather like a firewall protects your computer from the viruses and malware that circulate on the internet. In 1909, the South African-born biochemist Edwin Goldmann injected a deep blue dye into a bunch of different animals from frogs to monkeys and found that it stained cells throughout the body, except in the brain and the spinal cord, which remained 'as white as snow'.

In the first few years after the discovery of the hypocretins, there were some studies suggesting that one of the hypocretins might be able to make it through the blood–brain barrier in appreciable amounts, but these have been difficult to replicate. The current consensus is that neither of the hypocretins can penetrate the blood–brain barrier in a meaningful manner.

I could of course just slip a needle through the blood–brain barrier and inject hypocretins directly into my cerebrospinal fluid. Just before his death in 1913, Goldmann did just this in a follow-up experiment, giving rabbits a lumbar puncture and injecting trypan blue into the spinal column. This time, the colour showed up in the cerebrospinal fluid and the brain but did not spread to the rest of the body. In 2000, researchers did something similar with hypocretins, injecting them into the brains of narcoleptic rats. The formerly sleepy rats instantly became a lot less so. So injection of hypocretins into the brain works, but as keen as I am to find a replacement for this messenger it's not practical to live with a permanent spinal tap.

There has been excitement about the possibility of sniffing a fine mist of hypocretins, with the proteins bypassing the blood–brain barrier altogether and travelling to the very centre of the brain via the olfactory nerve. But as the amino acid sequence for the hypocretins is in the public domain there is nothing to protect with a patent, and so little reason for a pharmaceutical company to invest in such an idea.

If hypocretins really can travel along the olfactory nerve and into the brain in significant amounts, there is conceivably a way this might be a commercially viable project. When Lasse Ødegaard's young son was diagnosed with narcolepsy and cataplexy following vaccination with Pandemrix in 2009, he was not happy with the available medications. 'GHB, amphetamine and antidepressants were not drugs I wanted to give to my then six-year-old son,' says Lasse, a lawyer living and working in Oslo, Norway. He got up to speed on the hypocretins and came across the research papers on nasal delivery. A bit more searching brought him to OptiNose, a company founded by fellow Norwegian Per Djupesland to exploit a nifty device Djupesland invented (and patented) that promises far more efficient nose-to-brain delivery of medication than a conventional nasal spray. Lasse gave Djupesland a call, a collaboration that led the Norwegian Research Council to ring-fence almost $2 million for research into the delivery of hypocretins using the OptiNose device. Although this gadget can deliver another protein – oxytocin – in appreciable amounts to the brain, it remains to be seen whether the same could work for the hypocretins.

The more conventional route would be to invent a molecule that isn't a protein (so won't be digested in the gut), which can find its way into the brain (through the blood–brain barrier) and has the perfect structure (to bind to the hypocretin receptors). But coming up with a compound capable of activating the hypocretin receptors is much harder than identifying a chewing-gum-style compound to jam them, as Merck has done in the case of the insomnia drug Belsomra. The technical challenge is 'incomparably' greater, says Masashi Yanagisawa, who led one of the two research groups that independently discovered the hypocretins in 1998.

This doesn't mean it's impossible. In 2017, Yanagisawa and colleagues published data on the most potent such compound to date, a small molecule called YNT-185. Injections of this molecule into narcoleptic mice significantly reduced cataplexy, improved wakefulness and cut down on the unusually high levels of REM that are typical of narcolepsy. The affinity of YNT-185 for the hypocretin receptor is still not enough to warrant a clinical trial, but Yanagisawa's group already has several potential other candidates under investigation. 'The best one is almost 1,000 times stronger than YNT-185,' he says.

There are many hurdles to clear before a molecule like this becomes a medicine, but with the insights we've gained into the functions of the hypocretins so far, there is reason to hope that one day – and not too far in the future – there will be safe and effective drugs that have been designed specifically to pick up the slack in some of the neurological networks that lie downstream from the hypocretins. The latest treatment for narcolepsy and cataplexy, for instance, is a drug called pitolisant (branded Wakix), which was created with a very specific neurological receptor and pathway in mind, one that goes quiet in the absence of hypocretins. If Yanagisawa and others are successful, there may even be drugs that actually mimic the hypocretins themselves.

As things stand, there is insufficient investment in this research and development, most likely because there is a perception that the market for such an innovation is small. I have a vested interest so I am clearly partial. But everything I know about narcolepsy and everything I've learned about the function of the hypocretins strongly suggests that these are exceedingly important molecules with an impressive diversity of functions, and that being able to amplify this system could have very many applications indeed.

The story of Belsomra illustrates this perfectly. Here is a drug designed specifically to block the hypocretin receptors with the explicit goal of helping to combat insomnia, but it is also being considered as a treatment to help shift-workers sleep during the hours of daylight, to improve the sleep of Alzheimer's patients, to combat drug addiction and to ease human panic disorder. By the same token, a drug that can act as a stand-in for the hypocretins is likely to serve a purpose way beyond the confines of narcolepsy. Something that tickled up the hypocretins would be useful for any condition where excessive daytime sleepiness is an issue, not to mention the myriad other situations where low levels of these messengers may play a role, including obesity, depression, post-traumatic stress disorder and dementia.

Then there is stem cell therapy. Sergiu Paşca has the office next to Emmanuel Mignot at Stanford. He can do clever things with cells, like taking a skin cell, rewinding it to an early developmental stage known as a pluripotent stem cell and then letting it mature into a brain cell. 'You can use this system to derive various brain regions and like a Lego game assemble them to form circuits in a dish,' he says. Recently, his lab has succeeded in taking a skin cell from someone with narcolepsy and transforming it into a neuron that manufactures hypocretins. Such cells are of huge experimental value. In particular, they could reveal the precise set of autoimmune events that leads to the destruction of the hypocretin neurons and hence narcolepsy, an insight that could reveal ways to prevent this kind of attack in the first place. In theory, it should also be possible to transplant such cells into the brains of people with narcolepsy and restore some of the function to their hypocretin system.

Before anyone gets too excited about what would effectively be a cure for narcolepsy, it's probably sensible to point out a few caveats. First, hypocretin neurons engineered through such biotechnological cunning are unlikely to be exactly the same as natural-born hypocretin cells. Second, this kind of operation – to insert cells into the brain – is not something to be taken lightly, as there's always a risk of doing more bad than good. Third, it's unlikely that a hypocretin neuron would ever grow projections to all the right places. Fourth, it's possible that the same autoimmune response that took out the hypocretin neurons in the first place could do the same again for the stem cells.

The story of narcolepsy and the hypocretins is of still wider significance to the entire field of sleep medicine. Nobody has yet come up with a satisfying neurological explanation for idiopathic hypersomnia or Kleine-Levin Syndrome, for instance. Similarly sleepwalking, sleep talking, REM sleep behaviour disorder, sleep paralysis, restless legs syndrome and periodic limb movement disorder. Without an understanding of the mechanism underlying these pathologies, the chances of effective treatment are much reduced. Yet each of these sleep disorders is likely to have its own hypocretin system, a dysfunctional pathway just waiting to be discovered. It is simply a question of resources and perseverance.

\* \* \*

While I wait for these developments, I have much to be getting on with. For the last two decades, I have taken my narcoleptic symptoms as a given, a set of problems over which I had little control. Now that I have learned more about how sleep works and the many connections between narcolepsy and other sleep disorders, I have come to realise that even without hypocretins there is a lot I can do to protect what sleep I do get and maybe even to improve it.

The circadian rhythm is ancient, far more ancient than sleep. In fact, a properly synched body clock is pretty much essential for good sleep. I'm rather ashamed to say that it is only recently that I have given my circadian biology a moment's thought. If I did at all, it was only to assume that it would be ticking along nicely in the background.

This is a poor assumption for many reasons. For a start, sleep disorders – by definition – interfere with the body's ability to get in synch, most obviously by altering the timing and exposure to sunlight. If the eyes are closed more often than they should be, it's just more likely that you are going to miss key photonic signals – the shadowy blues before dawn and after dusk and the bright luminescence of day. If this happens, the master clock, the suprachiasmatic nucleus, or SCN, will not be keeping time as well as it should be, and cells all round the body will be getting mixed messages about what to do when.

There are likely to be other considerations too. In the case of narcolepsy, for instance, there are direct neurological connections between the SCN and the hypocretin cells in the hypothalamus, and the SCN may be carrying out some of its rhythmic magic by driving daily and seasonal fluctuations in the expression of the hypocretins. There is also a distinct possibility that the communication goes the other way too, with hypocretins affecting the SCN itself. It is therefore perfectly possible that there is a circadian component to narcolepsy, with the SCN master clock ticking more softly than it should be and more often out of time than is desirable. Given this, it's all the more important for people with narcolepsy to pay attention to light, getting as much natural light as possible during the day and avoiding all bright lights in the evening. Sticking to the same bedtime and wake time (whether it's a weekday or a weekend) should help. With a strict routine, it's possible that the excessive daytime sleepiness will become more predictable, with attacks more like scheduled events and less likely to strike at whim.

Similarly, the risk of sleep-disordered breathing that comes with narcolepsy can be managed too. The changes in activity, metabolism and feeding behaviour that follow a loss of hypocretins very often cause an increase in body mass, and this in turn makes snoring more likely. It might not sound like a big deal but snoring will disturb the already fractured night-time sleep of the narcoleptic. The most obvious solution is to lose weight by upping levels of physical activity and cutting back on calorific intake. Throat exercises can't hurt, whether executed with the aid of a didgeridoo or just by performing some basic guttural training over the course of a day. It's a good idea to sleep on your side rather than on your back. A mouthguard that fixes your lower jaw forward by a few millimetres can help. If you are serious about stopping snoring, then cutting back on cigarettes and alcohol is absolutely essential.

Snoring is also a significant risk factor for sleep apnea, and it's important to figure out whether this is an issue. It would be easy, after all, to assume that the bouts of excessive sleepiness during the day were all down to the narcolepsy when some might be the result of an undiagnosed, but easily reversed apnea. A simple step to gauge your risk of sleep apnea is to perform the anaesthetist's 'STOP-Bang' assessment (see chapter 6). If you answer 'yes' to three or more of the eight questions, then you should probably raise the possibility with your doctor. Consider also one of any number of apps to record the noises you make during the night (I found Snore-Lab particularly well-designed). Improving sleep apnea, either by any of the anti-snoring measures above or by deploying a CPAP machine, is likely to result in untold improvements to sleep, quite literally overnight.

Sleep paralysis is one of the most common parasomnias that accompanies narcolepsy. These waking dreams need to be taken seriously, partly because they are so horrific but also because of what they do to sleep architecture. A tried-and-tested solution is to sleep on your side rather than on your back, but this can't ever stop this problem completely. The only way I have found to combat sleep paralysis is through medication, taking a small dose of the antidepressant clomi-pramine every night. Other medications are available, so if the first-in-line treatment doesn't help control these night-mares it's important to work – in collaboration with your GP and specialist – to find something that does. A drug that controls sleep paralysis will likely also reduce the high incidence of REM that is characteristic of narcolepsy. This might mean losing the unbelievably vivid, fabulously inventive dreams, but it will also reduce the number of times these events interfere with the deep, non-REM sleep that seems to be so important for the brain to function at its best during the daytime.

People with narcolepsy are also at greater risk of experiencing many of the other parasomnias, like REM sleep behaviour disorder and periodic limb movement disorder, which both take a toll on the quality of sleep. As with sleep paralysis, there are medications that can help with each of these conditions, so recognising them and seeking proper medical help could result in a very simple win.

When it comes to the fractured night-time sleep that hounds most people with narcolepsy, the absence of hypocretins almost certainly plays a part. It's perfectly possible, however, that the 3Ps – the predisposing, precipitating and perpetuating factors – might come into play too. In my own case, I believe that my chaotic scheduling, haphazard napping and poor sleep hygiene compounded the problem. It is probably not necessary to go on a course of CBT for insomnia to identify many of the factors that contribute to the broken sleep of narcolepsy. Simple improvements to sleep hygiene, such as working to improve the circadian signals (see above), avoiding caffeine-based drinks after midday, giving up on nicotine and reducing alcohol intake, are all sensible first steps. Colin Espie's book *Overcoming Insomnia and Sleep Problems* is packed full of all the evidence-based tricks to combat wakefulness at night, and following through with a course of Sleepio, effectively the online version of CBT-i, could be beneficial.

For people with narcolepsy the impacts of being deprived of quality sleep are all too obvious. There are cognitive failings, alarming in their extent and frequency. There are evident physiological ramifications. Psychological consequences are often harder to identify, but they are almost always present and can be seriously disabling. Maximising deep sleep and minimising interruptive intrusions of REM or PMLD are crucial, which is why a drug like sodium oxybate is so beneficial. Being supported by family and friends is important too. In the absence of such understanding, sharing experiences with others with first-hand experience online can be an important source of validation and advice.

I feared that paying attention to all these aspects of my sleep would be a chore, not to mention pointless. In fact, I have improved my sleep so much and the results are so evidently beneficial that working to improve my sleep has become a healthy, even enjoyable obsession.

When I set out to write this book, I really didn't know where I would end up. Lost in the narcoleptic wilderness, it was not easy to see the wider landscape of sleep. For every sleep disorder, the terrain is fraught with danger, and the route to higher ground and good sleep is hard to spot and even more perilous to follow. But quite unexpectedly, I have emerged on an outcrop, a vantage point from which I have a better perspective on the land of bad sleep.

When I think about trying to identify which of the many changes I've implemented might be the most important for the improvements I've seen in my sleep, I reach an interesting conclusion. Sleep is so fundamental to our existence and so easily affected by every single aspect of our lives that a single, simple remedy is unlikely to do that much good. The real gains are to be had by paying attention to everything and simultaneously. Those who take good sleep for granted are fortunate indeed. But with hard work across all the domains I've touched on in this book, those with bad sleep can end up in a better place too.

I have been surprised by my progress. I know that I will slip back down the scree slope I've climbed from time to time. With a sleep disorder like narcolepsy, which can be pinpointed to a spot in the centre of our being, this much is inevitable. But having made it thus far once, I am confident I can find my way out again, at least to where I am now and maybe even higher. The mountain I'm climbing doesn't take me to the fertile plain I once inhabited, the land of good sleep. There is a chasm that I cannot cross at present, but I live in hope that someone, someday will build me a bridge.

# Author's note

Narcolepsy is a wildly variable spectrum disorder, as indeed are most sleep disorders. This book is based in part upon my experience of living with narcolepsy, but also, importantly, on hundreds of interviews that I have conducted over the last five years with scientists, doctors and others like me who suffer from some kind of dysfunctional sleep. I have grounded these professional and personal insights in material from historical archives, academic publications and the mainstream media. Many of those suffering from sleep disorders whom I spoke to were willing to be referred to by name. Where I have changed names and other details to protect the identity of those concerned, I have done my utmost to reflect the spirit in which the interview was conducted. I am a trustee of Narcolepsy UK, but all views contained in this book are entirely my own and do not necessarily represent those of the charity.

In these pages, I have referred a lot to the hypocretins, a pair of related neuropeptides whose absence from the brain results in narcolepsy. In chapter 5, I have relayed the story of their discovery in the late 1990s, work that was carried out simultaneously by two entirely independent research groups. Neither knew that the other was working on the same problem. The first group to describe these proteins called them 'hypocretins'. The second group, publishing very shortly afterwards, named the same compounds 'orexins'. The decision of what to call them remains unsettled. In a book like this, it would be unthinkable to refer to 'the hypocretins/orexins', which is what happens in much of the scientific literature. My decision to go with 'the hypocretins' is in no way meant to marginalise the work of those in 'the orexins' camp. I do so for the following reasons. First, there is the question of precedence (the hypocretin paper came before the orexin paper). Second, the name 'orexins' implies they have a role in feeding (while they do, the more important function is in driving wakefulness). Third, I just prefer the word hypocretin (pronounced hype-oh-creetin).

I hope very much that readers will find much they can relate to in these pages. Sleep, however, is so endlessly complex a subject and one whose significance we are only just beginning to comprehend, it is inevitable there will be important points that have gone unsaid. If and when these occur to you, please get in touch by 'liking' the book on Facebook and leaving a post on the wall (SleepyheadByHenryNicholls), by messaging me on Twitter (@WayOfThePanda) or by sending me an old-school email to henry@henrynicholls.com. I want to know what you liked about this book and what you felt I missed. It will help me in my ongoing journey towards a greater understanding of sleep.

# Further Reading

When I thumb through my copy of Nathaniel Kleitman's *Sleep and Wakefulness*, I am in awe. This book was first published in 1939 and was decades ahead of its time. The revised and enlarged edition, and the one that I own, appeared in 1963. It contains 4,337 references and a comprehensive index. Without word processors, reference managers or the internet, I have absolutely no idea how this feat was even possible. Although much of the literature is out of date, this remains an encyclopaedic survey of the early literature on sleep and sleep disorders.

For a history of sleep medicine, one in which Kleitman plays a significant part, Kenton Kroker's *The Sleep of Others* is the definitive tome. It charts the transformation of sleep from a private to a public concern. The invention of the electroencephalogram, which opened an objective window on the sleeping brain, is a key moment in this journey towards the discovery of sleep apnea and the emergence of sleep disorders' medicine in the late twentieth century.

In a more popular vein, William Dement's *The Promise of Sleep* stands out as an entertaining and informative look at sleep from a pioneer of sleep medicine. More recently, there have been several excellent popular books on sleep, including *The Mind at Night* by Andrea Rock, *Night School* by Richard Weismann, *The Mystery of Sleep* by Meir Kryger and *Why We Sleep* by Matthew Walker.

When it comes to specific aspects of sleep and sleep disorders, there is plenty of further reading to be had. *A Woman in the Polar Night* by Christiane Ritter is a little known book about a year spent among fur-trappers in the Arctic and contains magical descriptions of light and darkness. Roger Ekirch's discovery of a biphasic pattern of pre-industrial sleep first appeared in his stunning article 'Sleep we have lost', *American Historical Review* 106.2, (2001), p. 343–86 and then in *At Day's Close: A History of Nighttime*. For more on the science of the circadian rhythm, see *Rhythms of Life* by circadian biology pioneer Russell Foster and look out for *Our Solar Bodies* by Linda Geddes.

When it comes to narcolepsy, it's well worth dipping into William Adie, 'Idiopathic narcolepsy: A disease *sui generis*; with remarks on the mechanism of sleep', *Brain* 49.3, (1926), 257–306 and Luman Daniels, 'Narcolepsy', *Medicine* 13.1, (1934), 1–122. Both these early theses are remarkable for how accurately they depict the narcoleptic condition. Julie Flygare's *Wide Awake and Dreaming* is a vivid record of the bewildering descent into narcolepsy and Claire Crisp's *Waking Mathilda* is an immensely powerful story of parents responding to the sudden onset of narcolepsy in their young daughter.

On sleep apnea there are any number of self-help books but a surprising dearth of popular science books on this subject. If you're in any doubt about the importance of breathing read Nick Lane's *Oxygen*.

If you want to get an idea what it might be like to experience sleep paralysis with accompanying hypnagogic hallucinations, you need look no further than Carla MacKinnon's short film *The Devil in the Room*. The shadows and the puppetry capture the horror perfectly. *The Terror That Comes in the Night* by David Hufford is a lovely book and the thesis that began to put these paranormal experiences on a more scientific footing. It is filled with transcripts from interviewees. Then there's the abundance of literature that features a bout of sleep paralysis. The most haunting of these, perhaps, is Guy de Maupassant's short story 'Le Horla'.

As with sleep apnea, the majority of books on insomnia are of the self-help variety. If you are struggling with this choice, the tome you need is *Overcoming Insomnia and Sleep Problems* by Colin Espie. Alternatively, sign up for an online course of *Sleepio* that will take you through the techniques that have the greatest demonstrable impact on insomnia.

Given that restless legs syndrome is so poorly understood and so readily ignored, it is perhaps not that surprising to find there are relatively few books on this condition. Karl-Axel Ekbom's original description of restless legs syndrome remains extremely powerful. See Karl-Axel Ekbom, 'Asthenia Crurum Paraesthetica («Irritable legs«)', *Acta Medica Scandinavica* 118.1–3, (1944), 197–209 and Ekbom, 'Restless Legs', *Acta Medica Scandinavica* 121.S158 (1945), 1–123.

If I were to recommend just one book though, it would be *Far From the Tree* by Andrew Solomon. It's a big book and (as far as I recall) contains nothing about sleep, but it's a stunning exploration of ability, disability, difference and identity. When I set out to write this book, I was captivated by the intelligent, compassionate way in which Solomon presented the stories of others and bound them up with his own. I tried to do something similar with *Sleepyhead*. If I have succeeded, even just occasionally, I'm happy. *Far From the Tree* remains a source of considerable inspiration.

# Acknowledgements

Thank you to my agent Will Francis and my editor Rebecca Gray for encouraging me to use narcolepsy as a starting point for an exploration of sleep disorders and seeing the project through to its conclusion. I learned a lot in the process and in so-doing arrived at a much richer understanding of narcolepsy than I could ever have imagined. Rebecca helped enormously as I battled with the sheer abundance of material and helped to chaperone it into a coherent structure. Jamie Keenan designed the great jacket and Bill Johncocks devised the excellent index. Thanks too to Penny Daniel, Andrew Franklin, Drew Jerrison, Ed Lake and everyone else at Profile Books who got behind this project.

I received a generous grant from The Society of Authors (The Author's Foundation) with which I was able to conduct several research trips that turned out to be central to this book. In 2016, I attended the annual meeting of the European Narcolepsy Network in Helsinki, meeting Markku Partinen, Rosa Peraita-Adrados, Massimo Zenti, Anders Lauszus and Lasse Ødegaard and many more besides. In 2017, I travelled to California to interview several sleep specialists and researchers at Stanford University. I owe a great debt to William Dement, Christian Guilleminault, Meir Kryger, Luis de Lecea, Jamie Zeitzer, Manuel Spitschan, Philippe Mourrain, Emmanuel Mignot and Watson the Chihuahua for giving up so much of their time. While I was in the Bay Area, I visited Samantha Lundquist of the Peninsula Humane Society & SPCA and her narcoleptic pit bull-Staffordshire-bulldog cross Charlie. Back in London, I attended a meeting of the Royal Society of Medicine, where I met Roger Ekirch, Francesco Cappuccio, Jim Horne and Dirk-Jan Dijk, all of whom were kind enough to offer further help in capturing their work accurately. Several of these interviews fed into a feature I wrote for Wellcome's publication mosaicscience.com. Parts of that story are republished here in Chapter 5 and Chapter 12, under a Creative Commons Licence. Thank you to Chrissie Giles for all her work.

I would like to thank all those other researchers who gave up their time to answer questions by email or over the phone, including Jerry Siegel, Gert Jan Lammers, Gert van Dijk, Vincenzo Donadio, Daniel Denis, Debra Skene, Jim Horne, Masashi Yanagisawa, Francesco Cappuccio, Ling Lin, Raffaele Ferri, Paul Coleman, Pierre-Hervé Luppi, Jamie Seymour, Lee Kavanau, Michael Mangan, Chris Idzikowski, Allan Cheyne, Gary Ashton-Jones, Karl Ekbom Jr, Michael Prerau, Jason Ong, Paul Reading, Daniel Denis, Hannah Wunsch, Jed Black, Gregory Stores, Adrian Williams, Sergiu Paşca and Per Djupesland. In piecing together episodes in the history of sleep research, I have been helped by correspondence and interviews with Kenton Kroker, Stephen Casper, Andrew Hogan, David Millet, Merrill Mitler and Armond Aserinsky. Peter Todd was generous in sharing his experience of representing those who developed narcolepsy as a result of vaccination with Pandemrix.

I have found great support and new friends among those with narcolepsy and their families. Thank you to all the staff and volunteers at Narcolepsy UK for their dedication to this important charity. I have never met anyone at a Narcolepsy UK meeting with whom I did not strike up an instant connection. Many people with narcolepsy are also to be found online, particularly on the very excellent Narcolepsy in the UK Facebook Group. There are many reasons to wish that social media had never been invented but this group (and others like it) are the considerable upside. I have had many constructive discussions with fellow narcoleptics that have helped enormously in my efforts to capture the infinite diversity of symptoms; in particular I would like to thank Matt O'Neill, Dorothy Ennis-Hand, Joan Arlott, Michelle Hicks, Tony Broad, and Josh and Caroline Hadfield for inviting me into their homes and agreeing to being named in the book. Thank you also to Ronald Embleton and family, Sarah Garvey, Rob Gold, Franck Bouyer, Sarah Jackson, Angela Wells, Emily Rounds, Zoe Shardlow, Sam Harthen, Joanna Nicholson, Lily Clarke, Pen Pearson, Trish Wood, Lucy Tonge, Dee Daud, Melissa Stenning, Chloe Glasson, Pamela Felton, Kris Jackson, Anne Nduati, Jane Wachera and Julie Flygare. There are so many more of you who were kind enough to share your experience of narcolepsy with me. Whether you appear in the book by name or not, I have learned something from every single one of you. It is important that we speak up to communicate the true impact of narcolepsy. Thank you all.

I found the similarities between narcolepsy and other sleep disorders far more significant than the differences. This made it easier than I'd imagined to empathise with those experiencing other conditions. In particular, I would like to thank Marta Bravo, Jonathan Patten, Dean Jordheim, Aaron Suarez, Kate Goodwin, Martin Creed, Paul Sandham, Deborah Henry-Adolph, Peter Hames, Paul Warren, Carla MacKinnnon, Jennifer Salinas and all those on my CBT-i course. Charlotte Hall suggested I look at the STOP-Bang questionnaire that anaesthetists use to assess the risk of sleep apnea.

Compared to most people with narcolepsy, my journey to diagnosis was short. For this, I would like to thank David Fish and colleagues at the National Hospital for Neurology and Neurosurgery in Queen Square in London. More recently, with the support of my GP Zsuzsanna Cassidy, I have come under the care of Hugh Selsick at Guy's Hospital Sleep Disorders Centre and David O'Regan at the Royal London Hospital for Integrated Medicine. Trying to manage the symptoms of narcolepsy, I now realise, is an ongoing project that requires considerable investment by both patient and clinician. If this collaboration is effective, I believe that in many cases it is possible to find new ways to improve life with this chronic condition.

I have lived with narcolepsy now for more than half my life and would like to acknowledge the support of my friends, often in a very real sense as they have propped me up in the throes of a cataplectic thrill. Thank you Zaid, Kate, Jimmy, Pia, Rufus, Melvin, George, Jason, Marisa, Matt, Sara, John, Gaia, Julian, Emma, Jeremy, Jessica, Marianne, Eugene, Adam, Rakie, Harry, Megan, Rachel, Jake, Saskia, Steve, Nens, Tony, Philippa, Kirstie, Hamish, Clare and the Celeriac XI. I owe so much to my family, Stella, John, Harry, Eddie, Tom, Ana, Pablo, Alvaro, Oliver, Hazel, Jim, Mary, Mark, Irma, Morris, but above all my wife Charlotte, without whom this would not have been possible.

# Notes

## 1 Bad sleep

*Nymphomania is the urge* Zoe Shardlow, 'Sleeping Around', in
  *Personal Experiences 2* (A UKAN publication, 1994), pp. 19–23.  p. 1
*five hours a day* Jerome M. Siegel, 'Clues to the Functions of
  Mammalian Sleep', *Nature*, 437.7063 (2005), 1264–71 <https://doi.
  org/10.1038/nature04285>.                                        p. 3
*20 hours out of 24* Henry Van Twyver and Truett Allison, 'Sleep in
  the Opossum Didelphis marsupialis', *Electroencephalography
  and Clinical Neurophysiology*, 29.2 (1970), 181–9 <https://doi.
  org/10.1016/0013-4694(70)90121-5>.                               p. 3
*beauty, weakness and strength cited in* Nathaniel Kleitman, *Sleep and
  Wakefulness* (University of Chicago Press, 1963), p. 3.          p. 4
*a 'defect', he argued* Wilson Philip, 'On the Nature of Sleep',
  *Philosophical Transactions* (1833), 72–87.                      p. 4
*four hours a night* Tom de Castella, 'Thatcher: Can People Get by on
  Four Hours' Sleep?', *BBC News*, 10 April 2013, <http://www.bbc.
  co.uk/news/magazine-22084671> [accessed 17 August 2017].   p. 5
*what's going on* Donald J. Trump, *Think Like a Billionaire* (Random
  House, 2004), p. xix; Lydia Ramsey, 'Why Trump May Only Need
  a Few Hours of Sleep Each Night', *Business Insider*, 8 December
  2016, <http://uk.businessinsider.com/some-people-only-need-a-
  few-hours-of-sleep-2016-12> [accessed 23 October 2017].      p. 5
*published in 1939* Kleitman, p. 4.                                p. 5
*biggest mistake* Allan Rechtschaffen, 'The Control of Sleep', in *Human
  Behavior and Its Control*, ed. by W. A. Hunt (Shenkman Publishing
  Company, 1971), 75–92.                                          p. 6

279

*increase in appetite* Stephanie M. Greer, Andrea N. Goldstein and
 Matthew P. Walker, 'The Impact of Sleep Deprivation on Food
 Desire in the Human Brain', *Nature Communications*, 4 (2013),
 2259 <https://doi.org/10.1038/ncomms3259>. p. 7

*risk of obesity* Sanjay R. Patel and Frank B. Hu, 'Short Sleep Duration
 and Weight Gain: A Systematic Review', *Obesity*, 16.3 (2008),
 643–53 <https://doi.org/10.1038/oby.2007.118>. p. 7

*high blood pressure* Lin Meng, Yang Zheng and Rutai Hui, 'The
 Relationship of Sleep Duration and Insomnia to Risk of
 Hypertension Incidence: A Meta-Analysis of Prospective Cohort
 Studies', *Hypertension Research*, 36.11 (2013), 985–95 <https://
 www.ncbi.nlm.nih.gov/pmc/articles/PMC3819519/> [accessed 8
 September 2016]. p. 7

*susceptibility to infection* Aric A. Prather and others, 'Behaviorally
 Assessed Sleep and Susceptibility to the Common Cold', *Sleep*,
 38.9 (2015), 1353–59 <https://doi.org/10.5665/sleep.4968>. p. 7

*likelihood of depression* Chiara Baglioni and others, 'Insomnia
 as a Predictor of Depression: A Meta-Analytic Evaluation of
 Longitudinal Epidemiological Studies', *Journal of Affective
 Disorders*, 135.1–3 (2011), 10–19 <https://doi.org/10.1016/j.
 jad.2011.01.011>. p. 7

*pace of cognitive decline* June C. Lo and others, 'Sleep Duration
 and Age-Related Changes in Brain Structure and Cognitive
 Performance', *Sleep*, 37.7 (2014), 1171–8 <https://doi.org/10.5665/
 sleep.3832>. p. 7

*into the twentieth century* Carlos H. Schenck, Claudio L. Bassetti
 and others, 'English Translations of the First Clinical Reports
 on Narcolepsy and Cataplexy by Westphal and Gélineau in the
 Late 19th Century, With Commentary', *Journal of Clinical Sleep
 Medicine*, 3.3 (2007), 301–11. p. 8

*just over 15 years* Dorothy Ennis-Hand, Interview with author, 16
 January 2013. p. 9

*got a proper diagnosis* Michelle Hicks, Interview with author, 26
 March 2014. p. 11

*patient health and well-being* Raymond C. Rosen and others, 'Physician Education in Sleep and Sleep Disorders: A National Survey of U.S. Medical Schools', *Sleep*, 16.3 (1993), 249–54.     p. 12

*a seven-year degree* Gregory Stores and C. Crawford, 'Medical Student Education in Sleep and Its Disorders', *Journal of the Royal College of Physicians of London*, 32.2 (1997), 149–53.     p. 12

*still needed all round* Gregory Stores, Email to author, 16 September 2016.     p. 12

*diagnosing the disorder* Russell Rosenberg and Ann Y. Kim, 'The AWAKEN Survey: Knowledge of Narcolepsy among Physicians and the General Population', *Postgraduate Medicine*, 126.1 (2014), 78–86 <https://doi.org/10.3810/pgm.2014.01.2727>.     p. 12

*mental health disorder like depression* Meir H. Kryger, Randy Walld, and Jure Manfreda, 'Diagnoses Received by Narcolepsy Patients in the Year Prior to Diagnosis by a Sleep Specialist', *Sleep*, 25.1 (2002), 36–41.     p. 13

*an average wait of around 15 years* Michael J. Thorpy and Ana C. Krieger, 'Delayed Diagnosis of Narcolepsy: Characterization and Impact', *Sleep Medicine*, 15.5 (2014), 502–7.     p. 14

*a remarkable 67 years* Gianina Luca and others, 'Clinical, Polysomnographic and Genome-Wide Association Analyses of Narcolepsy with Cataplexy: A European Narcolepsy Network Study', *Journal of Sleep Research*, 22.5 (2013), 482–95 <https://doi.org/10.1111/jsr.12044>.     p. 14

*on a surfboard* Narcolepsy in the UK, 'What's the Strangest Place You've Fallen Asleep?', 1 June 2017.     p. 15

## 2 Let there be light

*to sense the Sun without ever seeing it* André Klarsfeld, 'At the Dawn of Chronobiology', 2013 <https://docs.google.com/viewer?a=v&pid=sites&srcid=ZGVmYXVsdGRvbWFpbnxrbGFyc2ZlbGRhbmRy ZXxneDooOTIyYzUwY2IyYWVjMjg2> [accessed 16 August 2017].     p. 17

*they called it* Period Ronald J. Konopka and Seymour Benzer, 'Clock Mutants of Drosophila melanogaster', *Proceedings of the National Academy of Sciences*, 68.9 (1971), 2112–16.      p. 18

*the circadian rhythm within each cell* 'The 2017 Nobel Prize in Physiology or Medicine - Press Release', nobelprize.org, 10 February 2017, <http://www.nobelprize.org/nobel_prizes/medicine/laureates/2017/press.html> [accessed 29 October 2017].      p. 19

*near-24-hour period* Martin R. Ralph and Michael Menaker, 'A Mutation of the Circadian System in Golden Hamsters', *Science*, 241.4870 (1988), 1225–27.      p. 20

*20-hour tune* Martin R. Ralph and others, 'Transplanted Suprachiasmatic Nucleus Determines Circadian Period', *Science*, 247.4945 (1990), 975–78 <https://doi.org/10.1126/science.2305266>.      p. 20

*central rhythm of the SCN* Jamie Zeitzer, Interview with author, 23 February 2017.      p. 21

*paper that described the condition* Elliot D. Weitzman and others, 'Delayed Sleep Phase Syndrome: A Chronobiological Disorder with Sleep-Onset Insomnia', *Archives of General Psychiatry*, 38.7 (1981), 737–746. Emily Sloan is not the patient's real name.      p. 24

*spraying lavender on your pillow* Marta Bravo, Interview with author, 4 September 2015.      p. 26

*avoid commitment altogether* Jonathan Patten, Interview with author, 11 February 2016.      p. 27

*the clock change* Christopher M. Barnes and David T. Wagner, 'Changing to Daylight Saving Time Cuts into Sleep and Increases Workplace Injuries', *The Journal of Applied Psychology*, 94.5 (2009), 1305–17 <https://doi.org/10.1037/a0015320>.      p. 27

*accidents in the workplace and on the roads* Barnes and Wagner; Stanley Coren, 'Daylight Saving Time and Traffic Accidents', *New England Journal of Medicine*, 334.14 (1996), 924–5 <https://doi.org/10.1056/NEJM199604043341416>.      p. 28

*daylight saving* Kyoungmin Cho, Christopher M. Barnes and
  Cristiano L. Guanara, 'Sleepy Punishers Are Harsh Punishers',
  *Psychological Science*, 28.2 (2017), 242–247 <https://doi.
  org/10.1177/0956797616678437>.                                      p. 28

*mild deprivation too* Anders Knutsson, 'Health Disorders of Shift
  Workers', *Occupational Medicine*, 53.2 (2003), 103–8.              p. 29

*mental and physical health* Josephine Arendt, 'Biological Rhythms
  During Residence in Polar Regions', *Chronobiology International*,
  29.4 (2012), 379–94 <https://doi.org/10.3109/07420528.2012.66899
  7>.                                                                 p. 30

*out of a tin* Meredith Hooper, 'The Enduring Eye: the Antarctic
  Legacy of Sir Ernest Shackleton and Frank Hurley,' Royal
  Geographical Society (with IBG), 21 November 2015 – 28 February
  2016.                                                               p. 30

*polar mentality* Christiane Ritter, *A Woman in the Polar Night*, reprint
  edition (Fairbanks: University of Alaska Press, 2010).              p. 30

*changes of the seasons* Emil Kraepelin, *Manic-Depressive Insanity and
  Paranoia* (E. & S. Livingstone, Edinburgh, 1921) <http://archive.
  org/details/manicdepressiveiookrae> [accessed 9 May 2017].    p. 31

*anyone else experiencing seasonal mood changes* Sandy Rovner,
  'Healthtalk: Seasons of the Psyche', *The Washington Post*, 6
  December 1981, p. E5.                                              p. 31

*SAD* Norman E. Rosenthal and others, 'Seasonal Affective Disorder:
  A Description of the Syndrome and Preliminary Findings with
  Light Therapy', *Archives of General Psychiatry*, 41.1 (1984),
  72–80.                                                             p. 31

*two clearly distinguishable chunks* Thomas A. Wehr, 'The Durations
  of Human Melatonin Secretion and Sleep Respond to Changes in
  Daylength (Photoperiod)', *The Journal of Clinical Endocrinology &
  Metabolism*, 73.6 (1991), 1276–80 <https://doi.org/10.1210/jcem-73-
  6-1276>.                                                           p. 32

*long-day/short-night mode* Thomas A. Wehr, 'A "Clock for All
  Seasons" in the Human Brain', in *Progress in Brain Research*,
  ed. by A. Kalsbeek and others (Elsevier, 1996), 321–42 <http://

www.sciencedirect.com/science/article/pii/S0079612308604161>
[accessed 14 November 2016].                                    p. 32

*Jane Rowth had a particular impact* Roger Ekirch, Interview with
author, 7 February 2017.                                        p. 33

*a couple of days later* A. Roger Ekirch, *At Day's Close: A History of
Nighttime* (Phoenix, 2006), p. 307.                            p. 33

*between light and light* The quotations from Nathaniel Hawthorne
and Robert Louis Stevenson appear in A. Roger Ekirch, 'Sleep We
Have Lost: Pre-Industrial Slumber in the British Isles', *American
Historical Review*, 106.2 (2001), 343–386.                     p. 34

*premier sommeil* A. Roger Ekirch, 'Sleep We Have Lost'.        p. 34

*getting out of bed* A. Roger Ekirch, 'A Social History of Sleep –
Looking Back to What Was "Normal Sleep"' (Royal Society of
Medicine, 2017).                                               p. 35

*a 24-hour period* National Sleep Foundation Recommends New
Sleep Times, 2 February 2015, <https://sleepfoundation.org/press-
release/national-sleep-foundation-recommends-new-sleep-times>
[accessed 23 October 2017].                                    p. 36

*shift in the body clock* Mary A. Carskadon, Cecilia Vieira and
Christine Acebo, 'Association between Puberty and Delayed Phase
Preference', *Sleep*, 16.3 (1993), 258–262.                    p. 37

*fat sand rat* Mary A. Carskadon, 'Sleep in Adolescents: The Perfect
Storm', *Pediatric Clinics of North America*, 58.3 (2011), 637–47
<https://doi.org/10.1016/j.pcl.2011.03.003>.                   p. 37

*get better grades* Pamela V. Thacher and Serge V. Onyper,
'Longitudinal Outcomes of Start Time Delay on Sleep, Behavior,
and Achievement in High School', *Sleep*, 39.2 (2016), 271–81
<https://doi.org/10.5665/sleep.5426>.                          p. 38

*the first few lessons* Anne C. Skeldon, Andrew J. K. Phillips and
Derk-Jan Dijk, 'The Effects of Self-Selected Light-Dark Cycles
and Social Constraints on Human Sleep and Circadian Timing:
A Modeling Approach', *Scientific Reports*, 7 (2017), <https://doi.
org/10.1038/srep45158>.                                        p. 38

*early hours of the morning* Kathryn J. Reid and others, 'Sleep: A Marker of Physical and Mental Health in the Elderly', *The American Journal of Geriatric Psychiatry*, 14.10 (2006), 860–6 <https://doi.org/10.1097/01.JGP.0000206164.56404.ba>.    p. 38

*when everyone was asleep* David R. Samson and others, 'Chronotype Variation Drives Night-Time Sentinel-like Behaviour in Hunter–gatherers', *Proceedings of the Royal Society of London B*, 284.1858 (2017), <https://doi.org/10.1098/rspb.2017.0967>.    p. 39

*delay or advance their body clocks* Anthony N. Van Den Pol, Vinh Cao and H. Craig Heller, 'Circadian System of Mice Integrates Brief Light Stimuli', *American Journal of Physiology. Regulatory, Integrative and Comparative Physiology*, 275.2 (1998), R654–7.    p. 41

*something similar happened in rats* Andreas Arvanitogiannis and Shimon Amir, 'Resetting the Rat Circadian Clock by Ultra-Short Light Flashes', *Neuroscience Letters*, 261.3 (1999), 159–62 <https://doi.org/10.1016/S0304-3940(99)00021-X>.    p. 41

*same effect in hamsters* Luis Vidal and Lawrence P. Morin, 'Absence of Normal Photic Integration in the Circadian Visual System: Response to Millisecond Light Flashes', *Journal of Neuroscience*, 27.13 (2007), 3375–82 <https://doi.org/10.1523/jneurosci.5496-06.2007>.    p. 41

*delayed the biological clock by almost an hour* Jamie M. Zeitzer and others, 'Response of the Human Circadian System to Millisecond Flashes of Light', *PLoS ONE*, 6.7 (2011), e22078 <https://doi.org/10.1371/journal.pone.0022078>.    p. 42

*thinking about light in a similar way* Derk-Jan Dijk, Interview with author, 10 March 2017.    p. 43

*patterns of expression* Guro F. Giskeødegård and others, 'Diurnal Rhythms in the Human Urine Metabolome During Sleep and Total Sleep Deprivation', *Scientific Reports* 5 (2015), <https://dx.doi.org/10.1038/srep14843>.    p. 43

*played a role* Maurice M. Ohayon and Cristina Milesi, 'Artificial Outdoor Night-time Lights Associate with Altered Sleep Behavior

in the American General Population', *Sleep*, 39.6 (2016), 1311–20
    <https://doi.org/10.5665/sleep.5860>.                                    p. 44
*misleading our brains and bodies* Anne-Marie Chang and others,
    'Evening Use of Light-Emitting EReaders Negatively Affects Sleep,
    Circadian Timing, and Next-Morning Alertness', *Proceedings of
    the National Academy of Sciences*, 112.4 (2015), 1232–7 <https://doi.
    org/10.1073/pnas.1418490112>.                                            p. 44


3 Weak with laughter
Sleep and Wakefulness Kleitman, p. 235.                                       p. 47
*my knee thing* Julie Flygare, *Wide Awake and Dreaming: A Memoir of
    Narcolepsy* (Arlington, VA: Mill Pond Swan Publishing, 2012). p. 47
*an impending collapse* Narcolepsy in the UK, 'Ideophones for
    Cataplexy', 20 June 2014.                                                p. 47
*as a dead body falls* Giuseppe Plazzi, 'Dante's Description of
    Narcolepsy', *Sleep Medicine*, 14.11 (2013), 1221–23 <https://doi.
    org/10.1016/j.sleep.2013.07.005>.                                        p. 48
*during an attack* Schenck, Bassetti, and others.                            p. 48
*collapsing in response to emotions* Schenck, Bassetti and others.          p. 49
*a sharp-minded remark* Sebastiaan Overeem and others, 'The
    Clinical Features of Cataplexy: A Questionnaire Study in
    Narcolepsy Patients with and without Hypocretin-1 Deficiency',
    *Sleep Medicine*, 12.1 (2011), 12–18 <https://doi.org/10.1016/j.
    sleep.2010.05.010>.                                                      p. 50
*the stupidest things* Sarah Garvey, Interview with author, 25 September
    2015.                                                                    p. 51
*very rare form of epileptic attack* 'Henry Nicholls' Medical Notes'
    (National Hospital for Neurology and Neurosurgery, 1994).    p. 53
*in the present* Sebastiaan Overeem, Gert Jan Lammers, and J. Gert
    van Dijk, 'Cataplexy: 'Tonic Immobility' Rather than 'REM-
    Sleep Atonia'?', *Sleep Medicine*, 3.6 (2002), 471–77 <https://doi.
    org/10.1016/S1389-9457(02)00037-0>.                                      p. 54
*one of the whalers* Roy Chapman Andrews, 'A Remarkable Case
    of External Hind Limbs in a Humpback Whale'. *American*

*Museum Novitates*, 9 (1921), 1–6 <http://digitallibrary.amnh.org/handle/2246/4849> [accessed 23 October 2017].                    p. 55

*70 million years* Matthew P. Harris and others, 'The Development of Archosaurian First-Generation Teeth in a Chicken Mutant', *Current Biology*, 16.4 (2006), 371–7 <https://doi.org/10.1016/j.cub.2005.12.047>.                    p. 55

*known or unknown conditions* Charles Darwin, *The Variation of Animals and Plants under Domestication*, 2nd edn (John Murray, 1875), <http://darwin-online.org.uk/content/frameset?keywords=the%20which%20visible%20obesides%20changes%20undergoes%20it&pageseq=52&itemID=F880.2&viewtype=text> [accessed 20 August 2017].                    p. 55

*relax its guard* J. Gert van Dijk, Interview with author, 10 October 2016.                    p. 56

*the consequence of terror* Jonathan Couch, *Illustrations of Instinct Deduced from the Habits of British Animals* (John Van Voorst, 1847) <http://archive.org/details/illustrationsinoocoucgoog> [accessed 6 July 2016].                    p. 56

*hated the smell* Matt O'Neill, Interview with author, 25 June 2015. p. 59

*never sleeps in these attacks* William J. Adie, 'Idiopathic Narcolepsy: A Disease Sui Generis; with Remarks on the Mechanism of Sleep', *Brain*, 49.3 (1926), 257–306. Adie's description of narcolepsy and cataplexy is one of the most accurate I have come across.     p. 60

*reflex epileptic attacks* 'Henry Nicholls' Medical Notes'.                    p. 60

*the humorous version* Sophie Schwartz and others, 'Abnormal Activity in Hypothalamus and Amygdala during Humour Processing in Human Narcolepsy with Cataplexy', *Brain*, 131.2 (2008), 514–22 <https://doi.org/10.1093/brain/awm292>.                    p. 63

*playing dead* Vincenzo Donadio, Email to author, 16 July 2016.   p. 66

*when the sleep comes he's there* Massimo Zenti, Interview with author, 19 March 2016.                    p. 66

## 4 Stages of sleep

*acted as the receiver* Theodore J. La Vaque, 'The History of EEG. Hans Berger: Psychophysiologist. A Historical Vignette', *Journal of Neurotherapy*, 3.2 (1999), 1–9. <http://dx.doi.org/10.1300/J184v03n02_01>     *p. 67*

*life-long career in psychophysics* David Millett, 'Hans Berger: From Psychic Energy to the EEG', *Perspectives in Biology and Medicine*, 44.4 (2001), 522–42 <https://doi.org/10.1353/pbm.2001.0070>.p. 67

*Berger published his discovery* Hans Berger, 'Über Das Elektrenkephalogramm Des Menschen', *European Archives of Psychiatry and Clinical Neuroscience*, 87.1 (1929), 527–570.     p. 68

*several different stages to sleep* Edgar D. Adrian and Bryan H. C. Matthews, 'The Interpretation of Potential Waves in the Cortex', *The Journal of Physiology*, 81.4 (1934), 440–71 <https://doi.org/10.1113/jphysiol.1934.sp003147>; Alfred L. Loomis, E. Newton Harvey and Garret Hobart, 'Potential Rhythms of the Cerebral Cortex During Sleep', *Science*, 81.2111 (1935), 597–98 <https://doi.org/10.1126/science.81.2111.597>; Alfred L. Loomis, E. Newton Harvey and Garret Hobart, 'Further Observations on the Potential Rhythms of the Cerebral Cortex During Sleep', *Science*, 82.2122 (1935), 198–200 <https://doi.org/10.1126/science.82.2122.198>.     p. 68

*not gradual but sudden* Robert W. Lawson, 'Blinking and Sleep', *Nature*, 165.4185 (1950), 81–82 <https://doi.org/10.1038/165081b0>.     p. 69

*the most distinguished sleep researcher in the world* Eugene Aserinsky, 'Memories of Famous Neuropsychologists: The Discovery of REM Sleep', *Journal of the History of the Neurosciences*, 5.3 (1996), 213–27.     p. 69

*golden manure* Aserinsky.     p. 69

*chicken walking through a barnyard* Armond Aserinsky, Interview with author, 22 June 2016.     p. 71

*acting out your dreams* Eugene Aserinsky and Nathaniel Kleitman, 'Regularly Occurring Periods of Eye Motility, and Concomitant Phenomena, during Sleep', *Science*, 118.3062 (1953), 273–274.     p. 72

*several years later* Michel Jouvet, F. Michael and J. Courjon, 'Sur Un Stade d'activité Électrique Cérébrale Rapide Au Cours Du Sommeil Physiologique', *Comptes Rendus Société Biologie*, 153 (1959), 1024–28.                                                        p. 72

*an objective way to study dreaming* William Dement and Nathaniel Kleitman, 'The Relation of Eye Movements during Sleep to Dream Activity: An Objective Method for the Study of Dreaming', *Journal of Experimental Psychology*, 53.5 (1957), 339–46.      p. 72

*when students fall asleep in class* William Dement, Interview with author, 19 February 2017.                                             p. 73

*a normal, healthy night's sleep* William Dement and Nathaniel Kleitman, 'Cyclic Variations in EEG during Sleep and Their Relation to Eye Movements, Body Motility, and Dreaming', *Electroencephalography and Clinical Neurophysiology*, 9.4 (1957), 673–90.                                                            p. 73

*harder to see* William C. Dement, *The Promise of Sleep: A Pioneer in Sleep Medicine Explores the Vital Connection Between Health, Happiness, and a Good Night's Sleep* (Macmillan, 2000), p. 20. p. 74

*observations of sleep without disturbing the sleeper* Dement, *The Promise of Sleep*, pp. 37–8.                                              p. 74

*all reported dreaming* William C. Dement, 'Dream Recall and Eye Movements during Sleep in Schizophrenics and Normals', *The Journal of Nervous and Mental Disease*, 122.3 (1955), 263–69.   p. 76

*proper functioning of the brain* William C. Dement, 'The Effect of Dream Deprivation', *Science*, 131.3415 (1960), 1705–7.          p. 76

*gleaned from dreams* Carey K. Morewedge and Michael I. Norton, 'When Dreaming Is Believing: The (Motivated) Interpretation of Dreams', *Journal of Personality and Social Psychology*, 96.2 (2009), 249–64 <https://doi.org/10.1037/a0013264>.               p. 77

*non-REM and REM* Michel Jouvet, 'Recherches Sur Les Structures Nerveuses et Les Mécanismes Responsables Des Différentes Phases Du Sommeil Physiologique', *Archives Italiennes de Biologie*, 100 (1962), 126–206. For a later review of Jouvet's early work, see Michel Jouvet, 'Biogenic Amines and the States of

Sleep', *Science*, 163.3862 (1969), 32–41 <https://doi.org/10.1126/
science.163.3862.32>. p. 77

*the occurrence and quality of dreams* Allan J. Hobson and Robert
McCarley, 'The Brain as a Dream State Generator: An Activation-
Synthesis Hypothesis of the Dream Process', *American Journal of
Psychiatry*, 134 (1977), 1335–1348. p. 78

*stained-glass window* Dement, *The Promise of Sleep*, p. 304. p. 78

*adaptive inactivity* Jerome M. Siegel, 'Sleep in Animals: A State of
Adaptive Inactivity', in *Principles and Practice of Sleep Medicine
(Fifth Edition)*, ed. by Meir H. Kryger, Thomas Roth and William
C. Dement (W. B. Saunders, 2011), pp. 126–38 <http://www.
sciencedirect.com/science/article/pii/B9781416066453000104>
[accessed 2 February 2015]. p. 79

*a piece of bread* Jerome M. Siegel, Interview with author, 11 May
2015. p. 80

*jellyfish* Jamie E. Seymour, Teresa J. Carrette and Paul A. Sutherland,
'Do Box Jellyfish Sleep at Night?', *Medical Journal of Australia*,
181.11/12 (2004), 706. p. 80

*distinct states of vigilance* J. Lee Kavanau, 'Vertebrates That Never
Sleep: Implications For Sleep's Basic Function', *Brain Research
Bulletin*, 46.4 (1998), 269–79 <https://doi.org/10.1016/S0361-
9230(98)00018-5>. p. 81

*demands on the brain* J. Lee Kavanau, 'Is Sleep's "Supreme Mystery"
Unraveling? An Evolutionary Analysis of Sleep Encounters No
Mystery; nor Does Life's Earliest Sleep, Recently Discovered
in Jellyfish', *Medical Hypotheses*, 66.1 (2006), 3–9 <https://doi.
org/10.1016/j.mehy.2005.08.036>. p. 81

*pointless information accumulated during the hours of wakefulness* John
W. Clark, Johann Rafelski and Jeffrey V. Winston, 'Brain without
Mind: Computer Simulation of Neural Networks with Modifiable
Neuronal Interactions', *Physics Reports*, 123.4 (1985), 215–273. p. 81

*neurological noise* Giulio Tononi and Chiara Cirelli, 'Sleep
and Synaptic Homeostasis: A Hypothesis', *Brain Research*

*Bulletin*, 62.2 (2003), 143–50 <https://doi.org/10.1016/j.
brainresbull.2003.09.004>.     p. 81

*might encode important memories* Luisa de Vivo and others,
'Ultrastructural Evidence for Synaptic Scaling across the Wake/
Sleep Cycle', *Science*, 355.6324 (2017), 507–10.     p. 82

*mice incapable of pruning* Graham H. Diering and others, 'Homer1a
Drives Homeostatic Scaling-down of Excitatory Synapses during
Sleep', *Science*, 355.6324 (2017), 511–15 <https://doi.org/10.1126/
science.aai8355>.     p. 82

*branched during non-REM sleep* Guang Yang and others, 'Sleep
Promotes Branch-Specific Formation of Dendritic Spines
after Learning', *Science*, 344.6188 (2014), 1173–78 <https://doi.
org/10.1126/science.1249098>.     p. 82

*stores of neurotransmitters and calcium* Vladyslav V. Vyazovskiy and
Kenneth D. Harris, 'Sleep and the Single Neuron: The Role
of Global Slow Oscillations in Individual Cell Rest', *Nature
Reviews Neuroscience*, 14.6 (2013), 443–51 <https://doi.org/10.1038/
nrn3494>.     p. 82

*toxic metabolic waste* Lulu Xie and others, 'Sleep Drives Metabolite
Clearance from the Adult Brain', *Science*, 342.6156 (2013), 373–77
<https://doi.org/10.1126/science.1241224>.     p. 82

*tests underpin the diagnosis of most sleep disorders* Dement,
Interview.     p. 82

*narcolepsy might be some kind of escape mechanism* Gerald Vogel,
'Studies in Psychophysiology of Dreams: III. The Dream of
Narcolepsy', *Archives of General Psychiatry*, 3.4 (1960), 421
<https://doi.org/10.1001/archpsyc.1960.01710040091011>.     p. 83

*may be a diagnostic aid* Allan Rechtschaffen and others,
'Nocturnal Sleep of Narcoleptics', *Electroencephalography and
Clinical Neurophysiology*, 15.4 (1963), 599–609 <https://doi.
org/10.1016/0013-4694(63)90032-4>.     p. 84

*hear Dement speak on sleep and dreams* Dement, Interview.     p. 85

*came up with the MSLT* Mary A. Carskadon, 'Guidelines for the Multiple Sleep Latency Test (MSLT): A Standard Measure of Sleepiness, Sleep 9.4 (1986), 519–524.                                  p. 85

*standard fare for narcoleptics* Michael Schredl, 'Dreams in Patients with Sleep Disorders', *Sleep Medicine Reviews*, 13.3 (2009), 215–21 <https://doi.org/10.1016/j.smrv.2008.06.002>.                 p. 87

*greatest clinical neurologist of all time* Macdonald Critchley, *Sir William Gowers, 1845–1915; a Biographical Appreciation.* (Heinemann, 1949).                                                  p. 87

*the diagnosis* Lina Nashef, *Gowers' Grand Round – 19/10/1995*, 19 October 1995.                                                  p. 87

*ten years ago* Emmanuel Mignot, 'Why We Sleep: The Temporal Organization of Recovery', *PLoS Biology*, 6.4 (2008), e106 <https://doi.org/10.1371/journal.pbio.0060106>.                  p. 88

*pretty close to the truth* Emmanuel Mignot, Interview with author, 20 February 2017.                                                  p. 88

*neurological development* Majid Mirmiran, 'The Function of Fetal/ Neonatal Rapid Eye Movement Sleep', *Behavioural Brain Research*, 69.1–2 (1995), 13–22 <https://doi.org/10.1016/0166- 4328(95)00019-P>.                                                  p. 89

*strengthening or pruning existing pathways* Francis Crick and Graeme Mitchison, 'REM Sleep and Neural Nets', *Behavioural Brain Research*, 69.1–2 (1995), 147–55 <https://doi.org/10.1016/0166- 4328(95)00006-F>; Wei Li and others, 'REM Sleep Selectively Prunes and Maintains New Synapses in Development and Learning', *Nature Neuroscience*, 20.3 (2017), 427–37 <https://doi. org/10.1038/nn.4479>.                                                  p. 89

*maintenance of threat-avoidance skills* Antti Revonsuo, 'The Reinterpretation of Dreams: An Evolutionary Hypothesis of the Function of Dreaming', *The Behavioral and Brain Sciences*, 23.6 (2000), 877–901.                                                  p. 90

*abstruse points of philosophy* Lyttleton Forbes Winslow, *On Obscure Diseases of the Brain and Disorders of the Mind: Their Incipient Symptoms, Pathology, Diagnosis, Treatment, and Prophylaxis*

(Blanchard and Lea, 1860) <https://archive.org/details/66330460R.
nlm.nih.gov> [accessed 23 October 2017].                                  p. 91

*Aserinsky and Kleitman* Philippe Mourrain, Interview with author, 22
February 2017.                                                                                p. 91

*REM in mammals* Lior Appelbaum and others, 'Sleep–wake
Regulation and Hypocretin–Melatonin Interaction in Zebrafish',
*Proceedings of the National Academy of Sciences*, 106.51 (2009),
21942–7 <https://doi.org/10.1073/pnas.906637106>.              p. 93

*mammalian sleep and REM* Mourrain, Interview.                      p. 94


## 5 Sleeping dogs don't lie

*a full-blown attack* Merrill M. Mitler and others, 'Narcolepsy-
Cataplexy in a Female Dog', *Experimental Neurology*, 45.2 (1974),
332–40 <https://doi.org/10.1016/0014-4886(74)90122-8>.      p. 96

*getting stranded* William C. Dement, Interview with author, 19
February 2017.                                                                                p. 96

*developing a cure* Mike Silverman, 'Poodle Has Sleeping Disease', *The
Courier-Express*, 22 May 1974, p. 6 <http://www.newspapers.com/
newspage/12598766/> [accessed 5 September 2014].            p. 97

*don't have an assistant* Dement, Interview.                              p. 97

*recessive gene* Arthur S. Foutz and others, 'Genetic Factors in Canine
Narcolepsy', *Sleep*, 1.4 (1979), 413–21.                              p. 98

*Mignot acknowledges* Mignot, Interview.                                  p. 99

*gut hormone* Luis de Lecea and others, 'The Hypocretins:
Hypothalamus-Specific Peptides with Neuroexcitatory Activity',
*Proceedings of the National Academy of Sciences*, 95.1 (1998),
322–7.                                                                                          p. 100

*mention of sleep* Takeshi Sakurai and others, 'Orexins and Orexin
Receptors: A Family of Hypothalamic Neuropeptides and G
Protein-Coupled Receptors That Regulate Feeding Behavior',
*Cell*, 92.4 (1998), 573–85 <https://doi.org/10.1016/S0092-
8674(00)80949-6>.                                                                    p. 100

*Most people said I was crazy* Mignot, Interview with author, 7 February 2012. The quotation from Ling Lin is based on Mignot's recollection of the conversation.     p. 100

*still in the running* In canine narcolepsy, the mutation responsible was discovered in the receptor for hypocretin 2. Both hypocretin 1 and hypocretin 2 can activate this receptor, but the activation of this pathway appears to be particularly important for driving wakefulness.     p. 101

*that August* Ling Lin and others, 'The Sleep Disorder Canine Narcolepsy Is Caused by a Mutation in The Hypocretin (Orexin) Receptor 2 Gene', *Cell*, 98.3 (1999), 365–376.     p. 102

*normally be active* Richard M. Chemelli and others, 'Narcolepsy in Orexin Knockout Mice: Molecular Genetics of Sleep Regulation', *Cell*, 98.4 (1999), 437–51 <https://doi.org/10.1016/S0092-8674(00)81973-X>.     p. 102

*trapped in a labyrinth* Christelle Peyron and others, 'Neurons Containing Hypocretin (Orexin) Project to Multiple Neuronal Systems', *Journal of Neuroscience*, 18.23 (1998), 9996–10015.     p. 104

*everything about them* Luis de Lecea, Interview with author, 21 February 2017.     p. 104

*switching on neuron after neuron after neuron* Matthew E. Carter and others, 'Mechanism for Hypocretin-Mediated Sleep-to-Wake Transitions', *Proceedings of the National Academy of Sciences*, 109.39 (2012), E2635–44 <https://doi.org/10.1073/pnas.1202526109>. p. 105

*being by a river* Wilder Penfield, 'Some Mechanisms Of Consciousness Discovered During Electrical Stimulation Of The Brain', *Proceedings of the National Academy of Sciences* 44.2 (1958), 51–66.     p .106

*silent during sleep* Maan Gee Lee, Oum K. Hassani and Barbara E. Jones, 'Discharge of Identified Orexin/Hypocretin Neurons across the Sleep-Waking Cycle', *Journal of Neuroscience*, 25.28 (2005), 6716–20 <https://doi.org/10.1523/jneurosci.1887-05.2005>.     p. 106

*envy the mouse* Antoine R. Adamantidis and others, 'Neural Substrates of Awakening Probed with Optogenetic Control of

Hypocretin Neurons', *Nature*, 450.7168 (2007), 420–4 <https://doi.org/10.1038/nature06310>. For the video clip of optogenetic activation of hypocretin neurons see <https://images.nature.com/full/nature-assets/nature/journal/v450/n7168/extref/nature06310-s2.mov>. p. 108

*surge of hypocretins* Ronald M. Salomon and others, 'Diurnal Variation of Cerebrospinal Fluid Hypocretin-1 (Orexin-A) Levels in Control and Depressed Subjects', *Biological Psychiatry*, 54.2 (2003), 96–104 <https://doi.org/10.1016/S0006-3223(02)01740-7>. p. 108

**6 Bad breath**

*cause of his moodiness* Alex Iranzo, Carlos H. Schenck and Jorge Fonte, 'REM Sleep Behavior Disorder and Other Sleep Disturbances in Disney Animated Films', *Sleep Medicine*, 8.5 (2007), 531–36 <https://doi.org/10.1016/j.sleep.2006.12.001>. p. 112

*5 in 10 women snore* Christopher Li and Victor Hoffstein, 'Snoring', in *Principles and Practice of Sleep Medicine (Fifth Edition)*, ed. by Meir H. Kryger, Thomas Roth and William C. Dement (W. B. Saunders, 2011), pp. 1172–82 <http://www.sciencedirect.com/science/article/pii/B9781416606645300102X> [accessed 26 February 2016]. p. 113

*many of his characters* Kerrie L. Schoffer and John D. O'Sullivan, 'Charles Dickens: The Man, Medicine, and Movement Disorders', *Journal of Clinical Neuroscience*, 13.9 (2006), 898–901 <https://doi.org/10.1016/j.jocn.2005.12.035>. p. 113

*tendency to sleep* Richard Caton, 'Case of Narcolepsy', *Transactions of the Clinical Society of London* 22 (1889), 133–7 <https://archive.org/stream/transactionscli48londgoog#page/n215/mode/2up> [accessed 12 February 2016]. p. 114

*the fat boy in Pickwick* Christopher Heath, 'Clinical Society of London', *British Medical Journal*, 1 (1889), 358. p. 114

*his business had gone bust* Caton. p. 114

*building up in his bloodstream* C. Sidney Burwell and others, 'Extreme Obesity Associated with Alveolar Hypoventilation – A Pickwickian

Syndrome', *The American Journal of Medicine*, 21.5 (1956), 811–18 <https://doi.org/10.1016/0002-9343(56)90094-8>. p. 115

*increasing somnolence* David Wang and others, 'Hypercapnia is a Key Correlate of EEG Activation and Daytime Sleepiness in Hypercapnic Sleep Disordered Breathing Patients', *Journal of Clinical Sleep Medicine* 10.5 (2014), 517–22 <https://doi.org/10.5664/jcsm.3700>. p. 115

*Henri Gastaut and his fellow medics* Henri Gastaut, C. A. Tassinari and B. Duron, 'Polygraphic Study of the Episodic Diurnal and Nocturnal (Hypnic and Respiratory) Manifestations of the Pickwick Syndrome', *Brain Research*, 1.2 (1966), 167–86 <https://doi.org/10.1016/0006-8993(66)90117-X>. p. 116

*symptoms during the daytime* Meir H. Kryger, Interview with author, 27 February 2016. p. 117

*the obese patient* Meir Kryger and others, 'The Sleep Deprivation Syndrome of the Obese Patient: A Problem of Periodic Nocturnal Upper Airway Obstruction', *The American Journal of Medicine*, 56.4 (1974), 531–39 <https://doi.org/10.1016/0002-9343(74)90485-9>. p. 117

*than anyone had hitherto imagined* Christian Guilleminault, Interview with author, 21 February 2017. p. 120

*one in 50 women* Naresh M. Punjabi, 'The Epidemiology of Adult Obstructive Sleep Apnea', *Proceedings of the American Thoracic Society*, 5.2 (2008), 136–143. p. 121

*someone with the condition* Kryger, Interview. p. 121

*choked himself to death* Meir H. Kryger, 'Sleep Apnea: From the Needles of Dionysius to Continuous Positive Airway Pressure', *Archives of Internal Medicine*, 143.12 (1983), 2301–3 <https://doi.org/10.1001/archinte.1983.00350120095020>. p. 122

*coronary heart disease* Khin Mae Hla and others, 'Coronary Heart Disease Incidence in Sleep Disordered Breathing'. *Sleep* 38.5 (2015), 677–684. p. 124

*those with no sleep apnea* Terry Young and others, 'Sleep Disordered
Breathing and Mortality: Eighteen-Year Follow-Up of the
Wisconsin Sleep Cohort', *Sleep* 31.8 (2008), 1071–8. P. 124

*body going limp* L.H. Stevens, 'Sudden Unexplained Death in Infancy:
Observations on an Natural Mechanism of Adoption of the Face
Down Position', *American Journal of Diseases and Children*, 110.3
(1965), 243–7.                                                    p. 125

*frequent apneas in adults* Richard L. Naeye, 'Pulmonary Arterial
Abnormalities in the Sudden-Infant-Death Syndrome', *New
England Journal of Medicine* 289.22 (1973), 1167–70; Adrian
Williams, G. Vawter and L. Reid, 'Increased Muscularity of
the Pulmonary Circulation in Victims of Sudden Infant Death
Syndrome', *Pediatrics* 63.1 (1979), 18–23.                      p. 125

*five times above normal* Javier F. Nieto and others, 'Sleep-Disordered
Breathing and Cancer Mortality', *American Journal of Respiratory
and Critical Care Medicine* 186.2 (2012), 190–4.                 p. 126

*risk that a cell will turn cancerous* 'Association between Obstructive
Sleep Apnea and Cancer Incidence in a Large Multicenter Spanish
Cohort', *American Journal of Respiratory and Critical Care Medicine*
187.1.                                                            p. 126

*twice as many accidents* A. T. Mulgrew and others, 'Risk and Severity
of Motor Vehicle Crashes in Patients with Obstructive Sleep
Apnoea/Hypopnoea', *Thorax* 63.6 (2008), 538–41.                 p. 126

*property damage and lost productivity* Alex Sassani and others,
'Reducing Motor-Vehicle Collisions, Costs, and Fatalities by
Treating Obstructive Sleep Apnea Syndrome', *Sleep* 27.3 (2004),
453–458.                                                          p. 126

*attentiveness throughout the medical care system* Mary P. McKay, 'Fatal
Consequences: Obstructive Sleep Apnea in a Train Engineer', *The
Annals of Family Medicine* 13.6 (2015), 583–6.                   p. 127

*issue of resistance* Guilleminault, Interview.                  p. 129

*upper airway resistance syndrome* Christian Guilleminault and others,
'A Cause of Excessive Daytime Sleepiness: The Upper Airway
Resistance Syndrome', *Chest*, 104.3 (1993), 781–787.           p. 129

*body mass index of 30 or more* The State of Food and Agriculture (Food and Agriculture Organisation of the United Nations, 2013). p. 129

*size of your airway* Guilleminault, Interview. p. 130

*the hippopotamus or the leviathan* Robert Macnish, *The Philosophy of Sleep* (W. R. M'Phun, 1836), p. 207–8 <http://archive.org/details/philosophyofslee00macn> [accessed 24 October 2017]. p. 130

*major metabolic changes* Eric J. Heckman and others, 'Metabolomics in Sleep Apnea', *American Journal of Respiratory and Critical Care Medicine*, 191 (2015), A2708. <https://doi.org/10.1164/ajrccm-conference.2015.191.1_MeetingAbstracts.A2708>. p. 130

*sleep homeostasis and metabolism* Carla S. Möller-Levet and others, 'Effects of Insufficient Sleep on Circadian Rhythmicity and Expression Amplitude of the Human Blood Transcriptome', *Proceedings of the National Academy of Sciences*, 110.12 (2013), E1132–41 <https://doi.org/10.1073/pnas.1217154110>. p. 130

*characteristic of obesity* Karine Spiegel and others, 'Effects of Poor and Short Sleep on Glucose Metabolism and Obesity Risk', *Nature Reviews. Endocrinology*, 5.5 (2009), 253–61 <https://doi.org/10.1038/nrendo.2009.23>. p. 130

*in the general population* Simon W. Kok and others, 'Hypocretin Deficiency in Narcoleptic Humans is Associated with Abdominal Obesity', *Obesity Research* 11.9 (2003), 1147–54. In this study, the prevalence of obesity in narcolepsy patients is 33 per cent compared to 12.5 per cent amongst controls. p. 130

*character from a kids' cartoon* Junko Hara and others, 'Genetic Ablation of Orexin Neurons in Mice Results in Narcolepsy, Hypophagia, and Obesity', *Neuron*, 30.2 (2001), 345–354. p. 131

*the hypocretin system* Antoine Adamantidis and Luis de Lecea, 'Sleep and Metabolism: Shared Circuits, New Connections', *Trends in Endocrinology & Metabolism*, 19.10 (2008), 362–70 <https://doi.org/10.1016/j.tem.2008.08.007>. p. 131

*further investigations* Mahesh Nagappa and others, 'Validation of the STOP-Bang Questionnaire as a Screening Tool for Obstructive

Sleep Apnea among Different Populations: A Systematic Review and Meta-Analysis.' *PLoS ONE* 10.12 (2015), e0143697 <https:// doi.org/10.1371/journal.pone.0143697>     p. 132

*increasing the risk of stroke* Jin-Gun Cho and others, 'Tissue Vibration Induces Carotid Artery Endothelial Dysfunction', *Sleep* 34.6 (2011), 751–7.     p. 133

*ear closest to the noise* Maya Sardesai, A.K. Tan and M. Fitzpatrick, 'Noise-Induced Hearing Loss in Snorers and Their Bed Partners', *Journal of Otolaryngology* 32.2 (2003), 141–5.     p. 133

*cured his apnea* Aron Suarez, Email to author, 7 December 2016.     p. 133

*having a didgeridoo* Milo A. Puhan and others, 'Didgeridoo Playing as Alternative Treatment for Obstructive Sleep Apnoea Syndrome: Randomised Controlled Trial', *British Medical Journal* 332.7536 (2006), 266–70 <https://doi.org/10.1136/ bmj.38705.470590.55>.     p. 134

*wakefulness during the day* Kátia C. Guimarães and others, 'Effects of Oropharyngeal Exercises on Patients with Moderate Obstructive Sleep Apnea Syndrome', *American Journal of Respiratory and Critical Care Medicine*, 179.10 (2009), 962–6 <https://doi. org/10.1164/rccm.200806-981OC>.     p. 134

*more active the following day* 'CPAP History: An Aussie Doctor and a Vacuum Cleaner', The Easy Blog (easybreathe.com, 24 September 2013).     p. 135

*able to watch television for several hours* Colin E. Sullivan and others, 'Reversal of Obstructive Sleep Apnoea by Continuous Positive Airway Pressure Applied Through the Nares', *The Lancet* 317.8225 (1981), 862–5.     p. 135

*no suitable masks* Scott T. Johnson and Jerry Halberstadt, *Phantom of the Night. Overcoming Sleep Apnea Syndrome and Snoring* (New Technology Pub, 1996).     p. 135

*very large corporation* Ibid.     p. 135

*life-threatening pulmonary and cardiovascular events* William C. Dement, 'My Nomination for the Nobel Prize in Physiology and

Medicine', Focus. *Journal for Respiratory Care & Sleep Medicine*,
July (2009). p. 136

*completely different reason* Kroker, *The Sleep of Others* (University of
Toronto Press, 2007). p. 136

*public object* Kenton Kroker, Interview with author, 22 July 2017. p. 136

*comes back immediately* Malcolm Kohler and others, 'Effects of
Continuous Positive Airway Pressure Therapy Withdrawal in
Patients with Obstructive Sleep Apnea', *Americal Journal of
Respiratory and Critical Care Medicine* 184.10 (2011), 1192–9. p. 136

*reducing the number of apneas* Patrick J. Strollo, Jr. and others, 'Upper-
Airway Stimulation for Obstructive Sleep Apnea', *New England
Journal of Medicine* 370.2 (2014), 139–49. p. 137

*wrote in the 1990s* in Johnson and Halberstadt, *Phantom of the
Night*. p. 137

### 7 The perfect neurological storm

*genetic mutation* Hyun Hor and others, 'A Missense Mutation in
Myelin Oligodendrocyte Glycoprotein as a Cause of Familial
Narcolepsy with Cataplexy', *The American Journal of Human
Genetics*, 89.3 (2011), 474–79 <https://doi.org/10.1016/j.
ajhg.2011.08.007>. p. 141

*remains unclear* Rosa Peraita-Adrados, Interview with author, 5
February 2016. p. 141

*performing ordinary chores* Schenck, Bassetti and others. p. 141

*less than 5 per cent of cases* Yves Dauvilliers, Isabelle Arnulf and
Emmanuel Mignot, 'Narcolepsy with Cataplexy', *The Lancet*
369.9560 (2007), 499–511. p. 141

*log fell on his head* Schenck, Bassetti and others. p. 142

*violent assaults or sports injuries* Hilaire J. Thompson, Wayne C.
McCormick and Sarah H. Kagan, 'Traumatic Brain Injury in
Older Adults: Epidemiology, Outcomes, and Future Implications',
*Journal of the American Geriatrics Society* 54.10 (2006),
1590–5. p. 142

*diagnosed as narcolepsy* Ennis-Hand, Interview. p. 142

*the responsible factor* A. Wilson Gill, 'Idiopathic and Traumatic
Narcolepsy', *The Lancet*, 237.6137 (1941), 474–476.                    p. 143

*chromosome six* Emmanuel Mignot and others, 'Narcolepsy and
Immunity', *Advances in Neuroimmunology*, 5.1 (1995), 23–37
<https://doi.org/10.1016/0960-5428(94)00043-N>.                    p. 143

*HLA in common* Mehdi Tafti and others, 'DQB1 Locus Alone
Explains Most of the Risk and Protection in Narcolepsy with
Cataplexy in Europe', *Sleep*, 37.1 (2014), 19–25 <https://doi.
org/10.5665/sleep.3300>. For patients with narcolepsy but
no cataplexy, there is also an increase in the prevalence of
DQB1*06:02, but the link is not so clear-cut.                    p. 144

*environmental triggers* Emmanuel Mignot, Interview with author,
7 February 2012.                    p. 144

*15 years old* Yves Dauvilliers and others, 'Age at Onset of Narcolepsy
in Two Large Populations of Patients in France and Quebec',
*Neurology*, 57.11 (2001), 2029–33 <https://doi.org/10.1212/
WNL.57.11.2029>.                    p. 144

*strep throat* Adi Aran and others, 'Elevated Anti-Streptococcal
Antibodies in Patients with Recent Narcolepsy Onset', *Sleep*, 32.8
(2009), 979–983.                    p. 144

*Spanish influenza* Dilip K. Gandhi and others, 'Narcolepsy: Is It a
Sequelae of Auto-Immune Encephalitis/Encephalitis Lethargica?',
*Journal of Neurology, Neurosurgery & Psychiatry*, 83.3 (2012), e1–e1
<https://doi.org/10.1136/jnnp-2011-301993.152>.                    p. 145

*every November there's a trough* Fang Han and others, 'Narcolepsy
Onset Is Seasonal and Increased Following the 2009 H1N1
Pandemic in China', *Annals of Neurology* 70.3 (2011), 410–17. p. 145

*84 deaths* 'Outbreak of Swine-Origin Influenza A (H1N1) Virus
Infection – Mexico, March–April 2009', Press Release from the
Center for Disease Control and Prevention, 30 April 2009 <http://
www.cdc.gov/mmwr/preview/mmwrhtml/mm58d0430a2.htm>
[accessed 24 October 2017].                    p. 147

*global pandemic* Declan Butler, 'Flu Pandemic Underway', *Nature
News*, 11 June 2009 <https://doi.org/10.1038/news.2009.564>. p. 147

*that was really the key* Markku Partinen, Interview with author, 5
    January 2016.                                                      p. 148
*near-permanent somnolence* Josh Hadfield and Caroline Hadfield,
    Interview with author, 14 August 2015.                            p. 150
*Pandemrix* Markku Partinen and others, 'Increased Incidence and
    Clinical Picture of Childhood Narcolepsy Following the 2009
    H1N1 Pandemic Vaccination Campaign in Finland', *PLoS ONE*,
    7.3 (2012), e33723 <https://doi.org/10.1371/journal.pone.
    0033723>.                                                         p. 150
*the pandemic was over* 'WHO Director-General Declares H1N1
    Pandemic Over', Press Release from the World Health
    Organisation, 10 August 2010 <http://www.euro.who.int/
    en/health-topics/communicable-diseases/influenza/news/
    news/2010/08/who-director-general-declares-h1n1-pandemic-over>
    [accessed 24 October 2017].                                       p. 150
*triggered by the swine flu vaccination* 'The MPA Investigates
    Reports of Narcolepsy in Patients Vaccinated with Pandemrix
    – Medical Products Agency, Sweden', Press Release from the
    Läkemedelsverket Medical Products Agency, 18 August 2010
    <https://lakemedelsverket.se/english/All-news/NYHETER-
    2010/The-MPA-investigates-reports-of-narcolepsy-in-patients-
    vaccinated-with-Pandemrix/> [accessed 24 October 2017].  p. 150–1
*narcolepsy in children* Hanna Nohynek and others, 'AS03 Adjuvanted
    AH1N1 Vaccine Associated with an Abrupt Increase in the
    Incidence of Childhood Narcolepsy in Finland', *PLoS ONE*, 7.3
    (2012), e33536 <https://doi.org/10.1371/journal.pone.0033536>.p. 151
*similar pattern* Amongst children and adolescents, vaccination with
    Pandemrix has been associated with an increase in the incidence
    of narcolepsy in all countries where the analysis has been carried
    out. In England, there was a 14-fold increase, Elizabeth Miller
    and others, 'Risk of Narcolepsy in Children and Young People
    Receiving AS03 Adjuvanted Pandemic A/H1N1 2009 Influenza
    Vaccine: Retrospective Analysis', *British Medical Journal*, 346.feb26
    2 (2013), f794–f794 <https://doi.org/10.1136/bmj.f794>; in Ireland

13-fold, National Narcolepsy Study Steering Committee and
others, 'Investigation of an Increase in the Incidence of Narcolepsy
in Children and Adolescents in 2009 and 2010', 2012 <http://www.
lenus.ie/hse/handle/10147/303432> [accessed 22 August 2017]; in
Denmark double, Leonoor Wijnans and others, 'The Incidence
of Narcolepsy in Europe: Before, during, and after the Influenza
A(H1N1)Pdm09 Pandemic and Vaccination Campaigns', *Vaccine*,
31.8 (2013), 1246–54 <https://doi.org/10.1016/j.vaccine.2012.12.015>;
in Sweden 7-fold, Medical Products Agency, Occurrence of
Narcolepsy with Cataplexy among Children and Adolescents in
Relation to the H1N1 Pandemic and Pandemrix Vaccinations –
Results of a Case Inventory Study by the MPA in Sweden during
2009–2010, 30 June 2011; in France 7-fold, Yves Dauvilliers and
others, 'Increased Risk of Narcolepsy in Children and Adults
after Pandemic H1N1 Vaccination in France', *Brain*, 136.8 (2013),
2486–96 <https://doi.org/10.1093/brain/awt187>. p. 151

*answered in the affirmative* Hadfield, Interview. p. 152

*compensate them and look after them* O'Neill, Interview. p. 152

*not exceeding 4 inches* International Labour Office, 'Compensation for
War Disabilities in Great Britain and the United States', *Studies
and Reports*, E.4 (1921), 1–85. p. 152

*72 per cent disabled* Anna Hodgekiss, 'Boy, 10, Who Developed
Narcolepsy after Swine Flu Jab is Awarded £120,000 in Damages,
*Mail Online*, 3 February 2016 <http://www.dailymail.co.uk/health/
article-3429659/Boy-10-developed-narcolepsy-swine-flu-jab-
awarded-120-000-damages.html> [accessed 23 October 2017]. p. 152

*successfully eradicated* Peter Todd, Email to author, 31 July 2017. p. 153

*H1N1 virus* Outi Vaarala and others, 'Antigenic Differences between
AS03 Adjuvanted Influenza A (H1N1) Pandemic Vaccines:
Implications for Pandemrix-Associated Narcolepsy Risk', *PLoS
ONE*, 9.12 (2014), e114361 <https://doi.org/10.1371/journal.
pone.0114361>. p. 154

*emotional stress* Cecilia Orellana and others, 'Life Events in the Year
Preceding the Onset of Narcolepsy', *Sleep* 17.8, Suppl (1994),
S50–3.                                                              p. 154

*immune defence* Emmanuel Mignot, Interview with author, 20
February 2017.                                                      p. 156

*other times of year* Yves Dauvilliers, Bertrand Carlander and others,
'Month of Birth as a Risk Factor for Narcolepsy', *Sleep*, 26.6
(2003), 663–666; Norbert Dahmen and Peter Tonn, 'Season of
Birth Effect in Narcolepsy', *Neurology*, 61.7 (2003), 1016–17; Dante
Picchioni, Emmanuel J. Mignot and John R. Harsh, 'The Month-
of-Birth Pattern in Narcolepsy is Moderated by Cataplexy Severity
and May be Independent of HLA-DQB1* 0602', *Sleep* 27.8
(2004), 1471–5.                                                     p. 156

*the stain doesn't stick at all* Christelle Peyron and others, 'A Mutation
in a Case of Early Onset Narcolepsy and a Generalised Absence
of Hypocretin Peptides in Human Narcoleptic Brains', *Nature
Medicine*, 6.9 (2000), 991–997,                                    p. 157

*hypocretin-producing cells* The estimates of the number of hypocretin
neurons vary, but this work suggests that in healthy controls there
are around 15,000 neurons on each side of the hypothalamus.
Christian R. Baumann and others, 'Loss of Hypocretin (Orexin)
Neurons with Traumatic Brain Injury', *Annals of Neurology*, 66.4
(2009), 555–59 <https://doi.org/10.1002/ana.21836>.               p. 157

*the neurotransmitter histamine* Joshi John and others, 'Greatly
Increased Numbers of Histamine Cells in Human Narcolepsy
With Cataplexy', *Annals of Neurology* 74.6 (2013), 786–93;
Philipp O. Valko and others, 'Increase of Histaminergic
Tuberomammillary Neurons in Narcolepsy', *Annals of Neurology*,
74.6 (2013), 794–804 <https://doi.org/10.1002/ana.24019>.   p. 158

**8 Lost in transition**

*thin veil descends* Christian Guilleminault, R. Phillips and William C.
Dement, 'A Syndrome of Hypersomnia with Automatic Behavior',

*Electroencephalography and Clinical Neurophysiology*, 38.4 (1975),
403–13. p. 160

*adding a crucifix* Shellay Maughan, 'Irish Horn Rosaries', *Irish Culture and Customs*, <http://www.irishcultureandcustoms.com/AEmblem/Rosaries.html> [accessed 24 October 2017]. p. 161

*recommended an exorcist* Zenti, Interview p. 161

*destroyed my career* Franck Bouyer, Interview with author, 21 July 2016. p. 162

*it turned up in the freezer* Sarah Jackson, Correspondence with author, 4 February 2016. p. 162

*bona fide eating disorder* Hal A. Droogleever Fortuyn and others, 'High Prevalence of Eating Disorders in Narcolepsy with Cataplexy: A Case-Control Study', *Sleep* 31.3 (2008), 335–41 <http://repository.ubn.ru.nl/handle/2066/73473> [accessed 24 October 2017]. p. 162

*prevalence of obesity* Ruth Janke van Holst and others, 'Aberrant Food Choices after Satiation in Human Orexin-Deficient Narcolepsy Type 1', *Sleep*, 39.11 (2016), 1951–59 <https://doi.org/10.5665/sleep.6222>. p. 163

*known what to do with* Angela Wells, Correspondence with author, 4 February 2016. p. 163

*shoplifting by the patient* Frank J. Zorick and others, 'Narcolepsy and Automatic Behavior: A Case Report', *Journal of Clinical Psychiatry*, 40.4 (1979), 194–97. Wendy Kaufman is an invented name based on the initials of the real patient. p. 164

*sleep talk every night* Rubens N. A. A. Reimão and Antonio B. Lefévre, 'Prevalence of Sleep-Talking in Childhood', *Brain and Development*, 2.4 (1980), 353–57 <https://doi.org/10.1016/S0387-7604(80)80047-7>. p. 165

*vivid dreams characteristic of REM* Arthur Arkin, 'Sleep-Talking: A Review', *Journal of Nervous and Mental Disease*, 143.2 (1966), 101–122, <http://journals.lww.com/jonmd/Fulltext/1966/08000/sleep_talking__a_review_.1.aspx> [accessed 23 August 2017]. p. 165

*more wicked, than I had imagined* George Orwell, *Such, Such Were the Joys* <https://www.amazon.co.uk/Such-Were-Joys-George-Orwell/dp/0141394374> [accessed 28 June 2017].　　　　　　　p. 166

*appearing to be incommoded* Sharda Umanath, Daniel Sarezky and Stanley Finger, 'Sleepwalking through History: Medicine, Arts, and Courts of Law', *Journal of the History of the Neurosciences*, 20.4 (2011), 253–76 <https://doi.org/10.1080/09647 04X.2010.513475>.　　　　　　　p. 167

*back in the 1990s* Carlos H. Schenck and Mark W. Mahowald, 'Review of Nocturnal Sleep-Related Eating Disorders', *International Journal of Eating Disorders*, 15.4 (1994), 343–356.　　　　p. 168

*such conviction is unlawful* Dement, *The Promise of Sleep*, p. 216.　　　　　　　p. 168

*lascivious cohabitation* Robert Wilhelm, 'The Sleepwalking Defense', *Murder By Gaslight*, 15 January 2011, <http://www.murderbygaslight.com/2011/01/sleepwalking-defense.html> [accessed 24 October 2017].　　　　　　　p. 169

*unqualified acquittal* Roger Broughton and others, 'Homicidal Somnambulism: A Case Report', *Sleep*, 17.3 (1994), 253–64.　p. 170

*slashing at wineskins* Alex Iranzo, Joan Santamaria and Martín de Riquer, 'Sleep and Sleep Disorders in Don Quixote', *Sleep Medicine*, 5.1 (2004), 97–100 <https://doi.org/10.1016/j.sleep.2003.05.001>.　　　　　　　p. 171

*what in heavens are you doing to me?* Carlos H. Schenck and others, 'Chronic Behavioral Disorders of Human REM Sleep: A New Category of Parasomnia', *Sleep*, 9.2 (1986), 293–308.　　p. 171

*RBD is certainly possible* Steven Morris, 'Devoted Husband who Strangled Wife in his Sleep Walks Free from Court', *The Guardian*, 20 November 2009,　　　　　　　p. 172

*without the muscle block* John Peever, Pierre-Hervé Luppi and Jacques Montplaisir, 'Breakdown in REM Sleep Circuitry Underlies REM Sleep Behavior Disorder', *Trends in Neurosciences*, 37.5 (2014), 279–88 <https://doi.org/10.1016/j.tins.2014.02.009>.　　　　p. 172

*neurodegenerative disorder* Alex Iranzo and others, 'Rapid-Eye-Movement Sleep Behaviour Disorder as an Early Marker for a Neurodegenerative Disorder: A Descriptive Study', *The Lancet Neurology*, 5.7 (2006), 572–77 <https://doi.org/10.1016/S1474-4422(06)70476-8>. p. 172

*onset of neurodegnerative disease* Carlos H. Schenck and Mark W. Mahowald, 'REM Sleep Behavior Disorder: Clinical, Developmental, and Neuroscience Perspectives 16 Years after Its Formal Identification in Sleep', *Sleep*, 25.2 (2002), 120–138. p. 173

*counteract it more easily* Pierre-Hervé Luppi, Interview with author, 23 March 2017. p. 173

*sexsomnia* Colin M. Shapiro, Nikola N. Trajanovic and J. Paul Fedoroff, 'Sexsomnia – a New Parasomnia?', *The Canadian Journal of Psychiatry*, 48.5 (2003), 311–317. p. 173

*indistinguishable from rape* Christian Guilleminault, Adam Moscovitch and others, 'Atypical Sexual Behavior during Sleep', *Psychosomatic Medicine*, 64.2 (2002), 328–336. p. 174

*plausible cases of sexsomnia* Michael A. Mangan and Ulf-Dietrich Reips, 'Sleep, Sex, and the Web: Surveying the Difficult-to-Reach Clinical Population Suffering from Sexsomnia', *Behavior Research Methods*, 39.2 (2007), 233–236. p. 174

*stops the sleep sex* Michael Mangan, *Sleepsex: Uncovered* (2001) <http://sleepsex.org/text/SleepsexUncovered.pdf> [accessed 21 October 2016]. p. 174

*broke down in court* 'Man Acquitted In Sex Assault Because he was Asleep', *CBC News*, 30 November 2005 <http://www.cbc.ca/news/canada/man-acquitted-in-sex-assault-because-he-was-asleep-1.568164> [accessed 24 October 2017]. p. 175

## 9 Ghosts and demons

*ridden by a hag* Owen Davies, 'The Nightmare Experience, Sleep Paralysis, and Witchcraft Accusations', *Folklore*, 114.2 (2003), 181–203. p. 179

*accounted for it* Henry Nicholls, Diary, 31 August 1994. p. 180

*terrifying dreams* John Bond, *An Essay on the Incubus, or Night-Mare* (D. Wilson and T. Durham, 1753) <http://archive.org/details/essayonincubusoroobond> [accessed 24 October 2017].　　p. 181

*attention from modern physicians* John Waller, *A Treatise on the Incubus, or Nightmare, Disturbed Sleep, Terrific Dreams, and Nocturnal Visions: With the Means of Removing These Distressing Complaints* (E. Cox and son, 1816).　　p. 181

*the imagination and the senses* Jules-Gabriel-François Baillarger, *Extrait d'un Mémoire Intitulé Des Hallucinations* (J. B. Baillière, 1846) <http://archive.org/details/extraitdunmmoireoobail> [accessed 24 October 2017].　　p. 181

*on the way into sleep* Alfred Maury, *Le Sommeil et les Rêves* (Didier, 1878) <http://archive.org/details/lesommeiletlesroimaurgoog> [accessed 24 October 2017].　　p. 181

*hypnopompic* Frederic W.H. Myers, Richard Hodgson and Alice Johnson, *Human Personality and Its Survival of Bodily Death* (Longmans, Green, 1903) <http://archive.org/details/humanpersonalityoimyer> [accessed 24 October 2017].　　p. 181

*bouts of sleep paralysis* Tomoka Takeuchi and others, 'Isolated Sleep Paralysis Elicited by Sleep Interruption', *Sleep*, 15.3 (1992), 217–25.　　p. 182

*plenty of other guises too* J. Allan Cheyne and Todd A. Girard, 'Paranoid Delusions and Threatening Hallucinations: A Prospective Study of Sleep Paralysis Experiences', *Consciousness and Cognition*, 16.4 (2007), 959–74 <https://doi.org/10.1016/j.concog.2007.01.002>.　　p. 182

*broken into her flat* Flygare.　　p. 182

*taken me hostage* Carla MacKinnon, 'The Sleep Paralysis Project,' electricsheepmagazine.co.uk <http://electricsheepmagazine.co.uk/features/2013/03/08/the-sleep-paralysis-project/>. For more about *The Devil in The Room* see <http://thesleepparalysisproject.org> [accessed 11 November 2017].　　p. 183

*contact with a ghost* David J. Hufford, *The Terror That Comes in the Night: An Experience-Centered Study of Supernatural Assault Traditions* (University of Pennsylvania Press, 1982), p. 186–7.   p. 185

*only certain people can see* Kris Jackson, Interview with author, 1 May 2014.   p. 185

*managed to escape* Erwin J. O. Kompanje, '"The Devil Lay upon Her and Held Her down" Hypnagogic Hallucinations and Sleep Paralysis Described by the Dutch Physician Isbrand van Diemerbroeck (1609–1674) in 1664', *Journal of Sleep Research*, 17.4 (2008), 464–67 <https://doi.org/10.1111/j.1365-2869.2008.00672.x>.   p. 185

*the rustling of feathers* Hufford, p. 224.   p. 185

*strung up by the neck* Davies.   p. 186

*neither speak nor stir* Davies.   p. 187

*the spectre came no more* Hufford, pp. 229–230.   p. 187

*I was paralysed* cited in Hufford, pp. 228.   p. 188

*incest wishes of infancy* Ernest Jones, *On The Nightmare* (Leonard and Virginia Woolf, 1931) <http://archive.org/details/onthenightmare032020mbp> [accessed 14 May 2014].   p. 188

*giving them no privacy* 'Michelle Hicks' Medical Notes' (Whittington Hospital, 1982).   p. 189

*accompanying hypnagogia* Jerome M. Schneck, 'Henry Fuseli, Nightmare, and Sleep Paralysis', *Journal of the American Medical Asssociation*, 207.4 (1969), 725–26 <https://doi.org/10.1001/jama.1969.03150170051011>.   p. 189

*her tender cries* Erasmus Darwin, *The Botanic Garden; a Poem, in Two Parts* (T. & J. Swords, 1807) <http://archive.org/details/botanicgardenpoeoodarw> [accessed 7 May 2014].   p. 190

*I am alone* Guy de Maupassant, 'Le Horla' (Paul Ollendorff, 1887), p. 10–11 <https://archive.org/details/lehorlaoomaupgoog> [accessed 25 October 2017].   p. 191

*a layer of powder* F. Scott Fitzgerald, *The Beautiful and the Damned* (C. Scribner's Sons, 1922), p. 242 <http://archive.org/details/beautifuldamnedoofitzrich> [accessed 25 October 2017].   p. 191

*sitting upon her chest as she lay* Thomas Hardy, *The Withered Arm* (Macmillan, 1888), p.44.      p. 191

*the horrid spell would be broken* Herman Melville, *Moby-Dick; or, The Whale* (Harpers and Brothers, 1851), p. 29.      p. 192

*he could not move or speak* Ernest Hemingway, *The Snows of Kilimanjaro, and Other Stories* (Charles Scribner's Sons, 1961), p. 192

*attempt to rise* Thomas De Quincey, *Confessions of an English Opium-Eater: Being an Extract from the Life of a Scholar* (William D. Ticknor, 1841).      p. 193

*one in three people* Brian A. Sharpless and Jacques P. Barber, 'Lifetime Prevalence Rates of Sleep Paralysis: A Systematic Review', *Sleep Medicine Reviews*, 15.5 (2011), 311–15 <https://doi.org/10.1016/j.smrv.2011.01.007>.      p. 193

*gargoyle type of creature* J. Allan Cheyne, Interview with author, 12 May 2014.      p. 193–4

*anecdotal wisdom* J. Allan Cheyne, 'Situational Factors Affecting Sleep Paralysis and Associated Hallucinations: Position and Timing Effects', *Journal of Sleep Research*, 11.2 (2002), 169–77.      p. 194

*avoid sleeping face up* Kompanje.      p. 195

*dilatable vessels of the Brain* Bond.      p. 195

*more benign category* Pauline Dodet and others, 'Lucid Dreaming in Narcolepsy', *Sleep*, 2014 <http://europepmc.org/abstract/med/25348131> [accessed 7 November 2014].      p. 195

*more readily than most* Dodet and others.      p. 195

*three figures* Richard J. McNally, 'Sleep Paralysis, Sexual Abuse, and Space Alien Abduction', *Transcultural Psychiatry*, 42.1 (2005), 113–22 <https://doi.org/10.1177/1363461505050715>.      p. 196

## 10 Wide awake

*very small portion of the night* Schenck, Bassetti and others.      p. 199

*half of those with narcolepsy* Giuseppe Plazzi, Leonardo Serra and Raffaele Ferri, 'Nocturnal Aspects of Narcolepsy with Cataplexy', *Sleep Medicine Reviews*, 12.2 (2008), 109–28 <https://doi.org/10.1016/j.smrv.2007.08.010>.      p. 199

*major new approach* Peter S. Kim, Annual Meeting of Stockholders, 25 May 2010. p. 200

*positive outcomes* W. Joseph Herring and others, 'Suvorexant in Patients with Insomnia: Pooled Analyses of Three-Month Data from Phase-3 Randomized Controlled Clinical Trials', *Journal of Clinical Sleep Medicine*, 12.09 (2016), 1215–25 <https://doi.org/10.5664/jcsm.6116>. p. 200

*most exciting thing* Paul Coleman, Interview with author, 22 June 2017. p. 201

*follow-up appointments* Hrayr P. Attarian, Stephen Duntley and Kelly M. Brown, 'Reverse Sleep State Misperception', *Sleep Medicine*, 5.3 (2004), 269–72 <https://doi.org/10.1016/j.sleep.2003.10.014>. p. 201

*she'd slept for almost seven hours* Jack Edinger and Andrew Krystal, 'Subtyping Primary Insomnia: Is Sleep State Misperception A Distinct Clinical Entity?', *Sleep Medicine Reviews* 7.3 (2003), 203–214. p. 202

*succession of intermediate states* Kleitman, p. 71. p. 203

*ball-squeezing* Michael J. Prerau and others, 'Tracking the Sleep Onset Process: An Empirical Model of Behavioral and Physiological Dynamics', *PLoS Computational Biology*, 10.10 (2014), e1003866 <https://doi.org/10.1371/journal.pcbi.1003866>. p. 203

*realms of possibility* Michael J. Prerau, Interview with author, 6 December 2016. p. 204

*cost to the National Health Service* Christina S. McCrae and others, 'Impact of Brief Cognitive Behavioral Treatment for Insomnia on Health Care Utilization and Costs', *Journal of Clinical Sleep Medicine*, 2014 <https://doi.org/10.5664/jcsm.3436>. p. 205

*great national embarrassment* David Colquhoun, 'Royal London Hospital Rebranded', DC's Improbable Science, 7 September 2010. p. 205

*sleep duration in the population* John A. Groeger, Fred R. H. Zijlstra and Derk-Jan Dijk, 'Sleep Quantity, Sleep Difficulties and Their Perceived Consequences in a Representative Sample of Some

2000 British Adults', *Journal of Sleep Research*, 13.4 (2004), 359–71 <https://doi.org/10.1111/j.1365-2869.2004.00418.x>. p. 206

*less than six nor more than nine* Anon, *Directions and Observations Relative to Food, Exercise and Sleep* (S. Bladon, 1772) <http://archive.org/details/2661933R.nlm.nih.gov> [accessed 25 October 2017]. p. 206

*80 per cent of people* Plazzi, Serra and Ferri. p. 207

*chronic insomnia* Arthur J. Spielman, 'Assessment of Insomnia', *Clinical Psychology Review*, 6.1 (1986). p. 207

*three female insomniacs for every two male insomniacs* Bin Zhang, Y. Wing and others, 'Sex Differences in Insomnia: A Meta-Analysis', *Sleep*, 29.1 (2006), 85. p. 208

*ethnicity may come into play* Kenneth L. Lichstein and others, 'Insomnia: Epidemiology and Risk Factors', in *Principles and Practice of Sleep Medicine (Fifth Edition)*, ed. by Meir H. Kryger, Thomas Roth and William C. Dement (W. B. Saunders, 2011), pp. 827–37 <http://www.sciencedirect.com/science/article/pii/B9781416066453000761> [accessed 5 February 2015]. p. 208

*family history of the condition* Yves Dauvilliers, Charles Morin and others, 'Family Studies in Insomnia', *Journal of Psychosomatic Research*, 58.3 (2005), 271–78 <https://doi.org/10.1016/j.jpsychores.2004.08.012>. p. 208

*awake for days* Jennifer Salinas, Interview with author, 6 February 2015. p. 208

*cells continue to chug along* Theresa E. Bjorness and Robert W. Greene, 'Adenosine and Sleep', *Current Neuropharmacology*, 7.3 (2009), 238–45 <https://doi.org/10.2174/157015909789152182>. p. 209

*40 minutes longer* Christopher L. Drake and others, 'Stress-Related Sleep Disturbance and Polysomnographic Response to Caffeine', *Sleep Medicine*, 7.7 (2006), 567–72 <https://doi.org/10.1016/j.sleep.2006.03.019>. p. 210

*equivalent amount of water* Kleitman, p. 61. p. 210

*more than half* Samuel Gresham and others, 'Alcohol and Caffeine: Effect on Inferred Visual Dreaming', *Science* 140 (1963), 1226–7.                                                                  p. 210

*heavier the smoker* Stefan Cohrs and others, 'Impaired Sleep Quality and Sleep Duration in Smokers – results from the German Multicenter Study on Nicotine Dependence', *Addiction Biology*, 19.3 (2014), 486–96 <https://doi.org/10.111 1/j.1369-1600.2012.00487.x>; Lin Zhang and others, 'Cigarette Smoking and Nocturnal Sleep Architecture', *American Journal of Epidemiology*, 164.6 (2006), 529–37 <https://doi.org/10.1093/aje/ kwj231>.                                                                  p. 210

*gastric reflux* William C. Orr and M.J. Harnish, 'Sleep-related Gastro-oesophageal reflux: provocation with a late evening meal and treatment with acid suppression', *Alimentary Pharmacology and Therapeutics*, 12.10 (1998), 1033–8.                                  p. 211

*injecting endorphins* Carl King and others, 'Effects of Beta-Endorphin and Morphine on the Sleep-Wakefulness Behavior of Cats', *Sleep*, 4.3 (1981), 259–62.                                                      p. 211

*cues for arousal* Richard R. Bootzin, Dana Epstein and James M. Wood, 'Stimulus Control Instructions', in *Case Studies in Insomnia*, ed. by Peter J. Hauri (Springer US, 1991), pp. 19–28 <https://doi.org/10.1007/978-1-4757-9586-8_2>.                    p. 212

*stimulus control* Charles M. Morin and others, 'Nonpharmacological Interventions for Insomnia: A Meta-Analysis of Treatment Efficacy', *American Journal of Psychiatry*, 151.8 (1994), 1172–80 <https://doi.org/10.1176/ajp.151.8.1172>.                              p. 212

*quickly flies away* Viktor E. Frankl, *The Doctor and the Soul* (Knopf, 1962).                                                              p. 213

*menopausal women* Kathryn A. Lee and Karen E. Moe, 'Menopause', in *Principles and Practice of Sleep Medicine (Fifth Edition)*, ed. by Meir H. Kryger, Thomas Roth and William C. Dement (W. B. Saunders, 2011) 1592–1601.                                          p. 215

*around one hour* Perry Nicassio and Richard Bootzin, 'A Comparison of Progressive Relaxation and Autogenic Training as Treatments

for Insomnia', *Journal of Abnormal Psychology*, 83.3 (1974), 253–60
<https://doi.org/10.1037/h0036729>.                                      p. 216

*taken out of the sails* Viktor E. Frankl, *Man's Search for Meaning*
(Beacon Press, 1992), p. 128. <http://archive.org/details/
MansSearchForMeaning-English> [accessed 25 October
2017].                                                                   p. 217

*not to think of a polar bear* Fyodor Dostoyevsky, *Winter Notes on
Summer Impressions* (Northwestern University Press, 1988).    p. 218

*fewer visits to the doctor* James W. Pennebaker and Sandra K.
Beall, 'Confronting a Traumatic Event: Toward an Understanding
of Inhibition and Disease', *Journal of Abnormal Psychology*,
95.3 (1986), 274–81 <https://doi.org/10.1037//0021-843X.
95.3.274>.                                                               p. 218

*physical and psychological health* Pasquale G. Frisna, Joan C. Borod,
and Stephen J. Lepore, 'A Meta-Analysis of the Effects of Written
Emotional Disclosure on the Health Outcomes of Clinical
Populations', *Journal of Nervous and Mental Disease*, 192.9 (2004),
629–34 <http://journals.lww.com/jonmd/Fulltext/2004/09000/A_
Meta_Analysis_of_the_Effects_of_Written.8.aspx> [accessed 28
January 2017].                                                           p. 219

*onset of slumber* Allison G. Harvey and Clare Farrell, 'The Efficacy
of a Pennebaker-Like Writing Intervention for Poor Sleepers',
*Behavioral Sleep Medicine*, 1.2 (2003), 115–24 <https://doi.
org/10.1207/S15402010BSM0102_4>.                                         p. 219

*articulatory suppression* A. B. Levey and others, 'Articulatory
Suppression and the Treatment of Insomnia', *Behaviour Research
and Therapy*, 29.1 (1991), 85–89 <https://doi.org/10.1016/S0005-
7967(09)80010-7>.                                                        p. 219

*crucial step* Peter Hames, Interview with author, 17 June 2015.    p. 222

*Sleepio works* Colin A. Espie and others, 'A Randomized, Placebo-
Controlled Trial of Online Cognitive Behavioral Therapy for
Chronic Insomnia Disorder Delivered via an Automated Media-
Rich Web Application', *Sleep*, 35.6 (2012), 769–81 <https://doi.
org/10.5665/sleep.1872>.                                                 p. 222

*affects sleep at night* Plazzi, Serra and Ferri.       p. 223

*histamine signalling system* Valko and others.       p. 224

## 11 Mind, body and soul

*death from want of sleep* Winslow, *On Obscure Diseases of the Brain and Disorders of the Mind*, p. 484.       p. 225

*proved the opposite* Dement, *The Promise of Sleep*, p. 242–7.       p. 226

*vital physiological function* Allan Rechtschaffen and others, 'Physiological Correlates of Prolonged Sleep Deprivation in Rats', *Science*, 221.4606 (1983), 182–84 <https://doi.org/10.1126/science.6857280>.       p. 226

*dead within months* D.T. Max, '3. Case Study: Fatal Familial Insomnia; Location: Venice, Italy; To Sleep No More', *The New York Times*, 6 May 2001 <http://www.nytimes.com/2001/05/06/magazine/3-case-study-fatal-familial-insomnia-location-venice-italy-to-sleep-no-more.html> [accessed 22 December 2016]; see also D.T. Max, *The Family That Couldn't Sleep: Unravelling a Venetian Medical Mystery* (Portobello Books, 2008).       p. 227

*degeneration of the thalamus* Glenn C. Telling and others, 'Evidence for the Conformation of the Pathologic Isoform of the Prion Protein Enciphering and Propagating Prion Diversity', *Science* 274.5295 (1996), 2079–82.       p. 227

*orchestration of sleep* Francesco Portaluppi and others, 'Progressive Disruption of the Circadian Rhythm of Melatonin in Fatal Familial Insomnia', *The Journal of Clinical Endocrinology and Metabolism*, 78.5 (1994), 1075–78 <https://doi.org/10.1210/jcem.78.5.8175963>.       p. 227

*TV in their room* Victoria Rideout, *Zero to Eight: Children's Media Use in America* (Common Sense Media, 2011).       p. 228

*screen-based media* Ofcom, 'Children and Parents: Media Use and Attitudes Report 2015', <https://www.ofcom.org.uk/research-and-data/media-literacy-research/children/children-parents-nov-15> [accessed 24 October 2017].       p. 228

*in front of a screen* Aric Sigman, 'Time for a View on Screen Time', *Archives of Disease in Childhood* 97 (2012), 935–42, <https://doi.org/10.1136/archdischild-2012-302196>. p. 228

*getting to sleep* Juulia E. Paavonen and others, 'TV Exposure Associated with Sleep Disturbances in 5- to 6-Year-Old Children', *Journal of Sleep Research*, 15.2 (2006), 154–61 <https://doi.org/10.1111/j.1365-2869.2006.00525.x>. p. 229

*diagnosed insomnia* Mari Hysing and others, 'Sleep and Use of Electronic Devices in Adolescence: Results from a Large Population-Based Study', *British Medical Journal Open*, 5.1 (2015), e006748 <https://doi.org/10.1136/bmjopen-2014-006748>. p. 229

*less than 85 per cent* Daniel L. King and others, 'The Impact of Prolonged Violent Video-Gaming on Adolescent Sleep: An Experimental Study', *Journal of Sleep Research*, 22.2 (2013), 137–43 <https://doi.org/10.1111/j.1365-2869.2012.01060.x>. The impact of screens on sleep is not just a problem for young children and adolescents. In adults too, more screen time tends to result in less sleep, either eating into the amount of time available for kipping or disrupting it. For example Jeroen Lakerveld and others, 'The Relation between Sleep Duration and Sedentary Behaviours in European Adults', *Obesity Reviews*, 17 (2016), 62–67 <https://doi.org/10.1111/obr.12381>. p. 229

*falling away from the bone* Deborah Henry-Adolph, Interview with author, 15 June 2016. p. 230

*on the map* Karl A. Ekbom, 'Asthenia Crurum Paraesthetica (Irritable Legs)', *Acta Medica Scandinavica*, 118.1–3 (1944), 197–209 <https://doi.org/10.1111/j.0954-6820.1944.tb17800.x>. p. 230

*Elvis legs* Richard P. Allen and others, 'Restless Legs Syndrome: Diagnostic Criteria, Special Considerations, and Epidemiology', *Sleep Medicine*, 4.2 (2003), 101–19 <https://doi.org/10.1016/S1389-9457(03)00010-8>. p. 231

*begin to weep* Karl A. Ekbom, 'Restless Legs', *Acta Medica Scandinavica*, 121.S158 (1945), 1–123 <https://doi.org/10.1111/j.0954-6820.1945.tb11978.x>. p. 231

*early hours* Claudia Trenkwalder and others, 'Circadian Rhythm of
   Periodic Limb Movements and Sensory Symptoms of Restless
   Legs Syndrome', *Movement Disorders*, 14.1 (1999), 102–10; Wayne
   A. Hening and others, 'Circadian Rhythm of Motor Restlessness
   and Sensory Symptoms in the Idiopathic Restless Legs Syndrome',
   *Sleep*, 22.7 (1999), 901–912; Jeanne F. Duffy and others, 'Periodic
   Limb Movements in Sleep Exhibit a Circadian Rhythm That Is
   Maximal in the Late Evening/Early Night', *Sleep Medicine*, 12.1
   (2011), 83–88 <https://doi.org/10.1016/j.sleep.2010.06.007>.   p. 231
*hypocretins in the evening* Paul Christian Baier, Robert Göder and
   Manfred Hallschmid, 'Circadian Variation of Hypocretin-1
   (Orexin A) in Restless Legs Syndrome', *Sleep Medicine*, 10.2
   (2009), 271 <https://doi.org/10.1016/j.sleep.2008.05.006>.   p. 231
*violent flailing* Jacques Montplaisir and others, 'Clinical,
   Polysomnographic, and Genetic Characteristics of Restless Legs
   Syndrome: A Study of 133 Patients Diagnosed with New Standard
   Criteria', *Movement Disorders*, 12.1 (1997), 61–5 <https://doi.
   org/10.1002/mds.870120111>.                                   p. 232
*rope around your spine* Henry-Adolph, Interview.                 p. 232
*symptoms to their GP* Richard P. Allen and others, 'Restless Legs
   Syndrome Prevalence and Impact. REST General Population
   Study', *Archives of Internal Medicine* 165.11 (2005), 1286–92.  p. 232
*one in 25 people* Maurice M. Ohayon and Thomas Roth, 'Prevalence
   of Restless Legs Syndrome and Periodic Limb Movement Disorder
   in the General Population', *Journal of Psychosomatic Research*, 53.1
   (2002), 547–54 <https://doi.org/10.1016/S0022-3999(02)00443-
   9>; Magdolna Hornyak and others, 'Periodic Leg Movements in
   Sleep and Periodic Limb Movement Disorder: Prevalence, Clinical
   Significance and Treatment', *Sleep Medicine Reviews*, 10.3 (2006),
   169–77 <https://doi.org/10.1016/j.smrv.2005.12.003>.          p. 233
*a big deal* Karl Ekbom, Jr, Interview with author, 6 June 2016.   p. 233
*exhaustion is overwhelming* Martin Creed, Interview with author,
   6 June 2016.                                                  p. 234

*diagnostic criteria for RLS* Raffaele Ferri and others, 'Different
   Periodicity and Time Structure of Leg Movements during Sleep in
   Narcolepsy/Cataplexy and Restless Legs Syndrome', *Sleep*, 29.12
   (2006), 1587.                                                      p. 235

*emotions must suffer* Michael O'Shea, 'Aspects of Mental Economy',
   *Bulletin of the University of Wisconsin*, 2.2 (1900), 33–100 <https://
   archive.org/details/aspectsmentaleco1shegoog> [accessed 25
   October 2017].                                                     p. 235

*burst out crying* Lily Clarke, Correspondence with author, 2 June
   2017.                                                              p. 236

*drawer is empty* Pen Pearson, Correspondence with author, 2 June
   2017.                                                              p. 236

*lose my car keys* Trish Wood, Correspondence with author, 1 June
   2016.                                                              p. 236

*short-term memory* Paula Alhola and Päivi Polo-Kantola, 'Sleep
   Deprivation: Impact on Cognitive Performance', *Neuropsychiatric
   Disease and Treatment*, 3 (2007), 553–67.                         p. 238

*full night's sleep* Nicholas J. Taffinder and others, 'Effect of Sleep
   Deprivation on Surgeons' Dexterity on Laparoscopy Simulator',
   *The Lancet*, 352.9135 (1998), 1191 <https://doi.org/10.1016/S0140-
   6736(98)00034-8>.                                                  p. 238

*shift-working* Presidential Commission on the Space Shuttle
   Challenger Accident, 6 June 1986, Appendix G <https://history.
   nasa.gov/rogersrep/genindex.htm> [accessed 25 October 2017];
   Merrill Mitler and others, 'Catastrophes, Sleep, and Public Policy:
   Consensus Report', *Sleep*, 11.1 (1988), 100–109.                 p. 238

*drifting off at the wheel* Marta Gonçalves and others, 'Sleepiness
   at the Wheel across Europe: A Survey of 19 Countries', *Journal
   of Sleep Research*, 24.3 (2015), 242–53 <https://doi.org/10.1111/
   jsr.12267>.                                                        p. 238

*legal limit for driving* Drew Dawson and Kathryn Reid, 'Fatigue,
   Alcohol and Performance Impairment', *Nature*, 388.6639 (1997),
   235–235 <https://doi.org/10.1038/40775>.                          p. 238

*obstructive sleep apnea* James A. Horne and Louise A. Reyner, 'Sleep
Related Vehicle Accidents', *British Medical Journal*, 310.6979
(1995), 565–67.                                                    p. 240

*normal levels* Charles F. P. George, 'Reduction in Motor Vehicle
Collisions Following Treatment of Sleep Apnoea with Nasal
CPAP', *Thorax* 56.7 (2001), 508–12 <http://dx.doi.org/10.1136/
thorax.56.7.508>.                                                  p. 240

*countermeasures* Louise Reyner and James A. Horne, 'Evaluation of
'in-Car' Countermeasures to Sleepiness: Cold Air and Radio',
*Sleep*, 21.1 (1998), 46–50.                                       p. 240

*refreshing* James Horne, Clare Anderson and Charlotte Platten, 'Sleep
Extension versus Nap or Coffee, within the Context of "Sleep
Debt"', *Journal of Sleep Research*, 17.4 (2008), 432–36 <https://doi.
org/10.1111/j.1365-2869.2008.00680.x>.                             p. 241

*self-imposed rules* Jim A. Horne, Interview with author, 10 February
2017.                                                              p. 242

*six hours sleep* Francesco P. Cappuccio and others, 'Meta-Analysis of
Short Sleep Duration and Obesity in Children and Adults', *Sleep*,
31.5 (2008), 619–26.                                               p. 242

*the short sleep* Philippa J. Carter and others, 'Longitudinal Analysis of
Sleep in Relation to BMI and Body Fat in Children: The FLAME
Study', *British Medical Journal*, 342 (2011), d2712 <https://doi.
org/10.1136/bmj.d2712>.                                            p. 243

*type 2 diabetes* H. Klar Yaggi, Andre B. Araujo and John B. McKinlay,
'Sleep Duration as a Risk Factor for the Development of Type 2
Diabetes', *Diabetes Care*, 29.3 (2006), 657–661.                  p. 243

*blood pressure* James E. Gangwisch and others, 'Short Sleep Duration
as a Risk Factor for Hypertension: Analyses of the First National
Health and Nutrition Examination Survey', *Hyptertension* 47.5
(2006), 833–9.                                                     p. 243

*blood vessels* Christopher Ryan King and others, 'Short Sleep Duration
and Incident Coronary Artery Calcification', *Journal of the
American Medical Association*, 300.24 (2008), 2859–66 <https://doi.
org/10.1001/jama.2008.867>.                                        p. 243

*stroke* Yue Leng and others, 'Sleep Duration and Risk of Fatal and Nonfatal Stroke: A Prospective Study and Meta-Analysis', *Neurology*, 84.11 (2015), 1072–9 <https://doi.org/10.1212/WNL.000000000001371>.                    p. 243

*coronary heart disease* Francesco P. Cappuccio and Michelle A. Miller, 'Is Prolonged Lack of Sleep Associated with Obesity?', *British Medical Journal*, 342. (2011), 3306 <https://doi.org/10.1136/bmj.d3306>.                    p. 243

*die early* Jane E. Ferrie and others, 'A Prospective Study of Change in Sleep Duration: Associations with Mortality in the Whitehall II Cohort', *Sleep*, 30.12 (2007), 1659; Francesco P. Cappuccio and others, 'Sleep Duration and All-Cause Mortality: A Systematic Review and Meta-Analysis of Prospective Studies', *Sleep*, 33.5 (2010), 585–92.                    p. 243

*moral judgements* William A. Broughton and Roger J. Broughton, 'Psychosocial Impact of Narcolepsy', *Sleep*, 17.8, Suppl (1994), S45–9.                    p. 243

*co-morbidities* Jed Black and others, 'Medical Comorbidity in Narcolepsy: Findings from the Burden of Narcolepsy Disease (BOND) Study', *Sleep Medicine* 33 (2017), 13–18.                    p. 244

*more than a decade* Ronald Embleton, Correspondence with author, 2012.                    p. 247

*the poor sleep* Robert E. Roberts and Hao T. Duong, 'The Prospective Association between Sleep Deprivation and Depression among Adolescents', *Sleep*, 37.2 (2014), 239–44 <https://doi.org/10.5665/sleep.3388>.                    p. 250

*buying clothing* Martin Creed, Email to author, 30 August 2016. p. 250

*moment of clarity* Martin Creed, Interview.                    p. 251

*I had whiplash* Julie Flygare, 'How Having Narcolepsy Messes With My Love Life,' *Women's Health*, 11 June 2015.                    p. 253

*orgasms* Dee-Dee, Correspondence with author, 24 January 2017.                    p. 253

*let go of the dream* Dee-Dee, Correpondence with author, 30 January 2017.                    p. 253

## 12 Good sleep

*less likely to occur* Emi Hasegawa and others, 'Serotonin Neurons in the Dorsal Raphe Mediate the Anticataplectic Action of Orexin Neurons by Reducing Amygdala Activity', *Proceedings of the National Academy of Sciences*, 114.17 (2017), E3526–35 <https://doi.org/10.1073/pnas.1614552114>.      p. 257

*too much during the day* Aatif M. Husain, Ruzica K. Ristanovic and Richard K. Bogan, 'Weight Loss in Narcolepsy Patients Treated with Sodium Oxybate', *Sleep Medicine*, 10.6 (2009), 661–63 <https://doi.org/10.1016/j.sleep.2008.05.012>.      p. 257

*Epworth Sleepiness Scale* 'About the ESS – Epworth Sleepiness Scale' <http://epworthsleepinessscale.com/about-the-ess/> [accessed 28 July 2017].      p. 257

*arousal and vigilance* M. A. Barrand, M. J. Dauncey and D. L. Ingram, 'Changes in Plasma Noradrenaline and Adrenaline Associated with Central and Peripheral Thermal Stimuli in the Pig', *The Journal of Physiology*, 316 (1981), 139–52.      p. 259

*white as snow* Marina Bentivoglio and Krister Kristensson, 'Tryps and Trips: Cell Trafficking Acros the 100-Year-Old Blood–Brain Barrier', *Trends in Neurosciences*, 37.6 (2014), 325–33.      p. 260

*appreciable amounts* Abba J. Kastin and Victoria Akerstrom, 'Orexin A but Not Orexin B Rapidly Enters Brain from Blood by Simple Diffusion', *Journal of Pharmacology and Experimental Therapeutics*, 289.1 (1999), 219–23.      p. 260

*a lot less so* Bentivoglio and Kristensson.      p. 260

*the olfactory nerve* Sam A. Deadwyler and others, 'Systemic and Nasal Delivery of Orexin-A (Hypocretin-1) Reduces the Effects of Sleep Deprivation on Cognitive Performance in Nonhuman Primates', *Journal of Neuroscience*, 27.52 (2007), 14239–47 <https://doi.org/10.1523/jneurosci.3878-07.2007>; Krister Kristensson, 'Microbes' Roadmap to Neurons', *Nature Reviews Neuroscience*, 12.6 (2011), 345–57 <https://doi.org/10.1038/nrn3029>.      p. 261

*OptiNose device* Lasse Ødegaard, Correspondence with author, 23 March 2016; 'Vil Lage Nesespray Mot Narkolepsi', *VG* <http://

www.vg.no/a/23325278> [accessed 28 July 2017]; Per Gisle
Djupesland and Arne Skretting, 'Nasal Deposition and Clearance
in Man: Comparison of a Bidirectional Powder Device and a
Traditional Liquid Spray Pump', *Journal of Aerosol Medicine
and Pulmonary Drug Delivery*, 25.5 (2012), 280–89 <https://doi.
org/10.1089/jamp.2011.0924>.                                    p. 261

*appreciable amounts* Per Gisle Djupesland, 'Nasal Drug Delivery
Devices: Characteristics and Performance in a Clinical Perspective
– a Review', *Drug Delivery and Translational Research*, 3.1 (2013),
42–62 <https://doi.org/10.1007/s13346-012-0108-9>.             p. 261

*discovered the hypocretins* Masashi Yanagisawa, Interview with author,
9 June 2017.                                                    p. 262

*sleep during the hours of daylight* 'Examination of the Effectiveness
of Suvorexant in Improving Daytime Sleep in Shift Workers'
<https://clinicaltrials.gov/ct2/show/NCT02491788?term=orexin&d
raw=1&rank=40> [accessed 30 June 2017].                        p. 263

*Alzheimer's patients* 'Safety and Efficacy of Suvorexant (MK-4305)
for the Treatment of Insomnia in Participants With Alzheimer's
Disease (MK-4305-061)' <https://www.clinicaltrials.gov/ct2/show/
NCT02750306?term=belsomra&draw=5&rank=2> [accessed 18
August 2017].                                                   p. 263

*combat drug addiction* 'Role for Hypocretin in Mediating Stress-
Induced Reinstatement of Cocaine-Seeking Behavior' <http://
www.pnas.org/content/102/52/19168.short> [accessed 17 February
2017].                                                         p. 263

*ease human panic disorder* 'The Role of Orexin in Human Panic
Disorder' <https://clinicaltrials.gov/ct2/show/NCT02593682>
[accessed 18 August 2017].                                     p. 263

*obesity* Judit A. Adam and others, 'Decreased Plasma Orexin-A Levels
in Obese Individuals', *International Journal of Obesity*, 26.2 (2002),
274.                                                            p. 263

*depression* Lena Brundin and others, 'Reduced Orexin Levels in the
Cerebrospinal Fluid of Suicidal Patients with Major Depressive

Disorder', *European Neuropsychopharmacology*, 17.9 (2007), 573–79
<https://doi.org/10.1016/j.euroneuro.2007.01.005>.                    p. 263
*post-traumatic stress disorder* Jeffrey R. Strawn and others, 'Low
    Cerebrospinal Fluid and Plasma Orexin-A (Hypocretin-1)
    Concentrations in Combat-Related Posttraumatic Stress Disorder',
    *Psychoneuroendocrinology*, 35.7 (2010), 1001–7 <https://doi.
    org/10.1016/j.psyneuen.2010.01.001>; África Flores and others,
    'Orexins and Fear: Implications for the Treatment of Anxiety
    Disorders', *Trends in Neurosciences*, 38.9 (2015), 550–59 <https://doi.
    org/10.1016/j.tins.2015.06.005>.                    p. 263
*dementia* Stephanie Lessig and others, 'Reduced Hypocretin (Orexin)
    Levels in Dementia with Lewy Bodies', *Neuroreport*, 21.11 (2010),
    756–60 <https://doi.org/10.1097/WNR.0b013e32833bfb7c>.  p. 263
*mature into a brain cell* Anca M. Pașca and others, 'Functional
    Cortical Neurons and Astrocytes from Human Pluripotent Stem
    Cells in 3D Culture', *Nature Methods*, 12.7 (2015), 671–78 <https://
    doi.org/10.1038/nmeth.3415>.                    p. 263
*derive various brain regions* Sergiu P. Pașca, Correspondence with
    author, 23 August 2017.                    p. 263
*seasonal fluctuations* Shengwen Zhang and others, 'Lesions of
    the Suprachiasmatic Nucleus Eliminate the Daily Rhythm of
    Hypocretin-1 Release', *Sleep*, 27.4 (2004), 619–627.      p. 265
*affecting the SCN* Lior Appelbaum and others, 'Sleep–wake
    Regulation and Hypocretin–melatonin Interaction in Zebrafish',
    *Proceedings of the National Academy of Sciences*, 106.51 (2009),
    21942–47 <https://doi.org/10.1073/pnas.906637106>.      p. 266

# Index

Note:

The index covers the main text but not the bibliography or notes. Inevitably, many entries begin with the word 'sleep'. The index is in word-by-word order, which means for example that 'sleep studies' appears before 'Sleepio' or 'Sleepsex'. The 3P Model of insomnia can be found at 'three', as though spelled out.

**Henry Nicholls** is a renowned science writer and author of three books. He has also written for *Nature, New Scientist*, and hosts the *Guardian*'s "Animal Magic" blog. Nicholls lives in London.